A TASTE OF NEW TESTAMENT GREEK

AN INTRODUCTION TO THE GRAMMAR OF THE GREEK NEW TESTAMENT

ROBERT RABE

ROBERTS INK

CONTENTS

A TASTE OF NEW TESTAMENT GREEK

An Introductory Course to the Language of the New Testament with Emphasis on Word Recognition

Robert Rabe
PhD Biblical Studies

Roberts Ink Press

Bremerton, Washington 98312

Contact: derseltenerabe@gmail.com

Manufactured in the United States of America

Formatted with Vellum

For every student desiring to draw closer to God through reading His Word in the original language.

ἐν τῇ καρδίᾳ μου ἔκρυψα τὰ λόγιά σου.
I have hidden Your Word in my heart.
Psalm 119:11 (Septuagint [Greek Old Testament])

PREFACE

ἡ βοήθειά μου παρὰ κυρίου.
My help comes from the Lord.
Psalm 121: 2a

Most students do not have the time to dedicate five or six years of their life to attend seminary with the goal of becoming a Doctor of New Testament Greek, but would be thrilled to be able to read the Greek text with relative ease. As explained later in the Introduction of this book, the goal and purpose of this endeavor is to help the student RECOGNIZE and COMPREHEND the text of the Greek New Testament.

This course does not purport to expose the student to every jot and tittle of the grammar of the Greek New Testament (GNT). Rather, it provides a *taste* of the language and a method through which the student can begin to comprehend the grammar and language of the GNT.

By definition, a taste of something does not involve consuming an entire meal, but merely sampling a particular food to determine whether one desires more. Another purpose of this book, then, is to provide a foretaste of things to come to help the student decide whether he or she desires to dive deeper into the study of the language. This volume contains most of the grammatical concepts the student would encounter if he or she read a lengthier publication dealing with the language of the GNT. However, such a publication could be between four hundred and seven hundred pages in length, such as "Basics of Biblical Greek," by William D. Mounce or "New Testament Greek for Beginners," by J. Gresham Machen. Many of the paradigms and vocabulary lists in Machen's classic textbook, which is now out of copyright, served as the inspiration for this book.

This volume contains grammatical explanations coupled with sample verses from the GNT as well as translation practice and grammar drills to help the student to begin to recognize the text. Answers to the translation exercises and drills are provided. All of the grammatical explanations and exercises and drills have been numbered so that the student can easily navigate the book. The book also offers a Greek Frequency Vocabulary, which lists nearly every Greek word occurring ten times or more in the GNT.

May I suggest an inexpensive Kindle edition of the GNT available on Amazon.com? The cost is only $7.00, but the book is easily worth one hundred times that amount. It is called, "The New Testament in the Original Greek: Enhanced Version with Parsing. Kindle Edition." Each word of the Greek text is a hyperlink to the lexicon. Simply click on a word in a passage and you will be taken to the lexicon where you will discover the meaning(s) of that particular Greek word as well as other valuable information about the term. Another click allows you to return to where you were in the text.

May God be with you on your journey.

Robert Rabe

INTRODUCTION

ἐν ταῖς ἐντολαῖς σου ἀδολεσχήσω.
I will meditate on your precepts.
Psalm 119:15a

Learning Objective

This course will empower the student to read the GNT with understanding and with a genuine sense of satisfaction, recognizing and comprehending the Greek text without memorizing long lists of Greek vocabulary.

Preliminary Comments

It is hoped that the study of New Testament Greek (NTG) will bring out clearly both the spiritual and the special educative value of the GNT. Direct personal contact with the Living Word of God and with so great a work of world literature must certainly prove a valuable personal and spiritual experience. As Raymond V. Schoder put it in his timeless textbook, "A Reading Course in Homeric Greek," "[while studying Greek] the student is sharpening his mind and sharpening his powers of thought and language by close working with the remarkably intellectual and highly organized Greek form of expression."

As indicated previously, this course is designed on the principle of observation, not memorization, on RECOGNIZING the meaning of the text, not on memorizing forms, on applying what has been learned to read the GNT, not on committing to memory (seemingly) endless lists of vocabulary words.

The course will familiarize the student with the words which MOST OFTEN occur in the GNT. As the great linguistic scholar William Sanford LaSor says in his "Handbook of New Testament Greek":

"The Learning of every Greek word as it is encountered – a practice used by many teachers of New Testament exegesis courses – is a horrible waste of time and effort. Of the 5,436 words in the vocabulary of the New Testament, 1,934 occur only one time each. That's 35% of the words! Very few of them are significant for exegetical purposes. 3,568 occur four times or less. That means that only 35% of the New Testament vocabulary (1,868 [words]) occurs five times or more. In teaching Greek we stress learning the most frequently used words ...With (a) basic vocabulary, a student can read quantities of the New Testament. We also encourage the student to *guess* the meaning of words in context. After all, the surest definition of any word will be gained by studying the word in the context in which it occurs."

Speakers of the English language can employ other tools in learning NTG: *familiarity and recognition*, two watchwords of this textbook. English-speaking students have practically been speaking Greek for years without ever realizing it. When we use words such as drama, photograph, pathos, politics, chorus, gymnasium and countless others (there are over half a million terms with Greek roots in your English dictionary!) we are speaking Greek under the pretense of English.

And yet another tool will be employed: learning to recognize the roots of words, and the meanings of these roots, as we find them embedded in the Greek words.

Schoder continues in this vein:

"'But is this practical?' you ask. In the highest sense, yes! It will not directly make you a living, nor teach you how to run a lathe, airliner or dictaphone. But it will teach you how to think, how to read with more understanding and speak with firmer mastery over words. It will enlarge your vocabulary and sharpen your grasp of the meaning of words. Since these are fundamental words which enter into all your life and every aspect of your job, Greek in developing them is more broadly practical than something which only teaches you how to operate a machine or add up company statistics ... Greek may not teach you a living, but it will teach you how to live. And if the purpose of education is to prepare you for living in vital contact with the best things in man's great inheritance, that you may get more out of life, then Greek is the most practical language in school. For that is what it specializes in."

Why Study New Testament Greek?

The New Testament has been translated dozens of times into English and into the other major languages of the earth. In addition, there are hundreds of commentaries which analyze and dissect every verse in the New Testament. So why would one invest the time and effort to learn NTG?

Dr. William D. Ramey, author of the online course "NT Greek In Session," answers this question.

"There are dozens of widely accepted English translations and a plethora of commentaries on every part of Scripture. So, why learn Greek? The numerous translations actually reveal the inherit dilemma. So many translations exist because the English language cannot render an exact translation from the Greek New Testament. First and foremost, the Greek language is highly inflected. Inflection allows an author to encode meaning – something that is lost in EVERY English translation. Furthermore, every Greek word and idiom has a semantic range of meaning, and that range in Greek can only be approximately represented in English.

"Any translation's stance on verbal inspiration is mirrored in the manner in which they translate. If a translation substitutes a noun for the original author's verb, an "a" for "the," replaces a subjunctive with an indicative, a past tense for a present, and so forth, then at best, the work has been seriously compromised, and at worst, the importance of verbal inspiration has been denied to the reader.

"Verbal inspiration extends only to the original wording and sentence order of Scripture. There is not a translation in existence that can replicate the original, no matter how widely accepted, revered, or good it appears to be. A. T. Robertson, a distinguished Greek scholar, wrote in his Preface to the third edition of *A Grammar of the Greek New Testament in the Light of Historical Research*, 'The Greek New Testament is *the* New Testament. All else is translation. Jesus speaks to us out of every page of the Greek. To get these words of Jesus it is worthwhile to plow through any grammar and to keep on to the end.'

"The study of the original languages may be viewed as the proper understanding of the biblical doctrine of inspiration. If we do indeed believe that every word in the original autograph was inspired by God the Holy Spirit, then should we not be diligently studying His message to us in its original form? If we are able to extract truth from God's Word that was previously indiscernible from a translation by learning Greek, then we shall succeed in our goal; not because of what has been translated, but because of what has been revealed from God! Furthermore, since the Greek New Testament is the Word of God, and the doctrine of biblical inspiration is what it is, then a thorough knowledge of biblical Greek is not a luxury, but indispensable."

A Guide to Learning Greek Vocabulary

Although most of us did not grow up immersed in the Greek language, thinking and speaking in Greek, we can at least learn to think in Greek passively. In order to think in this manner, we must rely on the author of the Greek text to take care in selecting the words and word endings for the student according to the Greek rules of grammar. He will.

All of us have many ideas, thoughts and concepts in common with the average Greek. We both think of such things as "eat," "clothes," "tree," etc. The Greeks merely employ a different symbol to represent things than we do. The basic concept, the basic idea is the same, but the words, the symbols, are different. Their word is just as reasonable as ours. You have grown up with a certain symbol for the concept "servant." You merely have to learn the Greek symbol for the same concept, which is **δοῦλος**.

Root and Stem Meanings

The way to solidify in your mind the Greek word **δοῦλος** is to do exactly the same thing you did when you learned the English word. When you first encountered the word "servant," it had little meaning to you. How did this word ultimately come to have meaning for you? You heard the word in context or looked up the meaning of the word in a dictionary and repeated the word to yourself while thinking of the object for which it stands. You formed a mental picture of a servant in your mind and consciously applied that image to this particular word. You sought over and over again to vividly link up in your mind's eye the image of a servant with the actual word. At one point you and I so closely associated in our minds the concept "servant" with our mental image of a servant that the two became forever inseparable. When you heard the word "servant," the mental image of a servant came rushing into your mind and when you needed to describe a person who serves another person you used the word "servant." That is how you learned English words and this is the way to learn Greek words as well.

Another tool the student can employ to solidify a Greek word in his or her consciousness is to take a piece of paper and to write the word over and over until it is fixed in the mind.

Because this textbook emphasizes RECOGNITION over memorization, many "Recognition Tips" will be employed. Here is the first one:

Recognition Tip

Purchase a deck of 3 X 5 cards. Write the Greek word or phrase on one side of a card and the English translation on the opposite side. Review the cards repeatedly until you recognize the Greek term and know its English meaning.

A Greek word, as you know or will know, consists of two parts: the stem of the word and the ending. You will soon learn that Greek is driven by the stems and the endings of words. As you familiarize yourself with Greek vocabulary, turn your attention to the stem of the word, that is, the first part of the word, and connect the stem with the idea or thing. (In our journey we will mention the stems of words many times.) When you become familiar with the endings of nouns, verbs, etc., memorize the force or function of that ending. The ending of a word provides it with a certain viewpoint or twist or taste, so to speak.

For instance, the genitive case (possessive case) ending of any Greek noun automatically gives that noun the meaning of "from" or "of." Add the genitive ending to the word "man" and that singular noun takes on the idea of "from the man" or "of the man." To impart some other meaning, viewpoint or twist to the word "man," we must add a different ending. For example, if we add the dative case ending in the singular, we get the idea of "to the man," or "for the man."

Thus, when the Greek noun **ἄνθρωπος**, which means "man," ends in **ου** (**ἀνθρώπου**), it becomes genitive singular and the meaning is, "OF the man" or "FROM the man." The ending, then, always indicates what particular aspect or twist that noun has. In this case, as stated, the meaning is "OF the man" or "FROM the man," as opposed to "TO the man" or "FOR the man."

The intelligent way to recognize (or even memorize) the **ου**-ending, then, is to hook it up in your mind with that particular aspect or use of all words: the ending **-ου** means OF, FROM. Then whenever you see it attached to, in this case, a certain noun stem, you will at once know what the author is saying about that idea; your mind will be focused on that one angle of the word's use. It will not have to run through the whole list of endings, parrot-like, one after the other until it comes to the one spelled **-ου**, then figure out what different force that ending gives the word from the other endings around it on any declension chart. (A declension chart for any given noun gives the various forms of that noun in the various grammatical cases.) When you see the ending **-ου**, your mind should automatically fall into one defined focus: OF or FROM whatever the writer is talking about.

Schoder has some thoughts along this line:

"The same with verb endings. Learn the function, the use of each (ending) in itself, whatever verb stem you find the ending on, (and) you will know that it means I AM NOW DOING IT, or THEY DID IT, etc. Stem and endings together will then function in their natural way and give a complete idea, for example, I-AM-NOW-THINKING, etc.

You will not understand all this theory the first time. Go over it again, trying to see the general point. Then try it out on the words you are learning day by day. After some experience of this sort, re-study the theory, and you will see much more clearly how it works; for you will find that this is what your mind has been doing all along – with some clumsiness at first, no doubt, but more easily and surely every day. By consciously combining theory and practice, and thinking this matter out for yourself as your experience grows, you will find that your skill in using Greek in in the right way, as a language, is improving constantly. The main thing is practice. The theory is only to set the practice going along the right lines, in the way your mind is made to work in learning languages."

ONE

Μακάριοι οἱ ἄμωμοι ἐν ὁδῷ, οἱ πορευόμενοι ἐν νόμῳ κυρίου.

Blessed are those whose ways are blameless, who walk in the law of the Lord.

Psalm 119:1

A Brief History of New Testament Greek

THE COUNTRY which we now call Greece consisted in the centuries before the birth of Christ of numerous separate city states. These states shared a common language and culture, although there were some linguistic and cultural differences from one area to another. Four main dialects existed – Attic and Ionic, which were related, and Aeolic and Doric. In the period before the birth of Christ, the dialect of Athens, Attic, became the principle literary form of the Greek language. Attic Greek is what is known today as "Classical Greek."

WITH THE CONQUESTS of the Macedonian warrior-king, Alexander the Great, Attic Greek became the official language of most of the Middle East, and was used as the language of learning, trade, and administration throughout the entire Macedonian Empire.

ALEXANDER THE GREAT (356 B.C. - 323 B.C.) was born in Pella, Macedonia, the son of Philip of Macedonia and his wife, Olympias, a Greek. Beginning at age 14, the young boy was tutored by the Greek philosopher and thinker, Aristotle, who had been summoned to Pella by Phillip. Under the tutelage of this Greek teacher, Alexander gained a love for all things Greek, especially philosophy and the tales of battle and conquest described in Homer's works. Technically, the Macedonians were not Greeks. The war-like Macedonians were distantly related to the Greeks, and under the leadership of Philip they had subjugated the country of Greece. Violence and intrigue were mainstays in the court of Philip and in the year 336 B.C. he was assassinated.

ALEXANDER, thus, ascended the throne at the age of 20. The young king conquered northern Africa, western Asia (to liberate the Greeks of Asia Minor) and lands as far east as India, until all of the known world was under his control. He died at age thirty-two in Babylon, a city he had hoped to make the capital of his empire.

Alexander's world conquest resulted in the Hellenization

of the ancient world. Greek culture and language changed the world, and in turn the world changed the Greek language. As we will discover below, not only did the various Greek dialects meld together, but the resulting language that became common to varied and diverse nationalities underwent a number of changes, some typical of what happens to languages over time, and some the result of the internationalization of the language.

HELLENISTIC GREEK IS the Greek that was common (**κοινός**) to the various peoples of the ancient world from approximately the 800 years between 300 B.C. and A.D. 500. Attic Greek, in the process of being transformed from the language of Plato and others to the common language of the people, became simpler grammatically and absorbed forms from other languages (including Latin). In the first century A.D., **κοινή** Greek was spoken and understood throughout Palestine and around the entire Mediterranean basin as well as in large portions of the Roman Empire and beyond. Rome was in control of the civilized world at the time of Christ and Latin was the official language of government. However, Greek remained the common language of communication and everyday concourse. One illustration of this point can be seen in John 19:20, where the inscription on the Cross was written in Aramaic (the local language), Latin (the official language of the Roman government), and Greek (the common language which almost everyone could speak and read).

Many scholars find it highly probable that Jesus gave much of his teaching in the Greek language, since He spoke to crowds containing people from throughout the Roman Empire, which was comprised of much of the territory that Alexander had conquered. It is probable that many portions of the Gospels contain the actual words of Christ rather than a translation from His native Aramaic to Greek.

We have already mentioned that during the classical period (from about 900 B.C. - 300 B.C.), the Greek language consisted of a number of dialects. Approximately 500 years before Christ, the Attic dialect attained the supremacy, especially as the language of Greek prose, literature and philosophy. This dialect became the language of Athens in her glory. It was the language of Thucydides, of Plato and of Demosthenes. Attic Greek was also the medium employed by most of the other great prose writers of Greece.

As Machen puts it in his introduction to *"New Testament Greek for Beginners"*:

"Various causes contributed to make the Attic dialect dominant in the Greek-speaking world. First and foremost must be put the genius of the Athenian writers. But the political and commercial importance of Athens was also not without its effect. Hosts of strangers came into contact with Athens through government, war and trade, and the Athenian colonies also extended the influence of the mother city. The Athenian Empire, indeed, soon fell to pieces. Athens was conquered first by Sparta in the Peloponnesian war, and then, in the middle of the fourth century before Christ, along with the other Greek cities, Athens came under the domination of the king of Macedonia, Philip. But the influence of the Attic dialect survived the loss of political power; the language of Athens also became the language of her conquerors."

As mentioned above, Macedonia was not originally a kingdom of the Greek Empire. The Macedonians adopted the dominant civilization of the day, which was the civilization of Athens. Philip's son, Alexander the Great, too, was enamored with the language and culture of Athens due largely to the influence of Aristotle. Once young Alexander ascended the throne of Macedonia, he rapidly made himself master of the whole eastern world, and the triumphs of Alexander and of the Macedonian army also became the triumphs of the Greek language in its Attic form. To quote Machen once again:

"The empire of Alexander, indeed, at once fell to pieces after his death in 323 B.C., but the kingdoms into which it was divided were, at least so far as the court and the governing classes were concerned, Greek kingdoms. Thus, the Macedonian conquests meant nothing less than the Hellenization of the East, or at any rate it meant an enormous acceleration of the Hellenizing process which had already begun."

In the two centuries prior to the birth of Christ, the Romans conquered the eastern part of the Mediterranean world. The conquers made no effort to suppress the Greek language, which through the influences of Alexander, had taken root there. Indeed, the conquerors to a very great extent were conquered by the culture which they had conquered. The Romans had already come under Greek influence, and they now utilized the Greek language in the administration of at least the eastern part of their vast empire. Thus, the language of the Roman Empire was not so much Latin as it was Greek.

The New Testament was penned during the **κοινή** period. Linguistically speaking, it is closely related to the Greek translation of the Old Testament called the "Septuagint," which was completed at the north African port of Alexandria in the centuries immediately preceding the Christian era. **κοινή** Greek is also closely related to certain other Christian writings of the early part of the second century after Christ, designated as the writings of the Apostolic Fathers. Within these three groups the language of the New Testament is easily supreme.

The language of the New Testament is related to, but does differ from the Greek of the great Attic prose writers such as Thucydides or Plato or Demosthenes. The centuries which intervened between the Attic writers and philosophers and the advent of New Testament Greek naturally brought with them linguistic changes. In addition, the **κοινή** dialect, in becoming a cosmopolitan world language, absorbed much foreign influence, including Semitic influence. **κοινή** Greek became the ideal vehicle for God's word to reach the greatest readership. It is significant that the Epistle to the Hebrews was written in Greek, not Aramaic; and the Epistle to the Romans was penned in Greek, not Latin.

The bottom line, according to Greek scholar William Mounce, is that "God used the common language to communicate the gospel. The gospel does not belong to the erudite alone; it belongs to all people. It now becomes our task to learn this marvelous language to help us make the grace of God known to all people."

TWO

ἐν ὅλῃ καρδίᾳ μου ἐξεζήτησά σε.
With my whole heart I sought you.
Psalm 119:10a

Greek Pronunciation

1. The Classical Greek Pronunciation scheme is presented below.

Shown, in order, are the English term for the Greek letter, the Greek term for the letter, the Greek lowercase letter, the Greek uppercase letter, and finally, how the letter is pronounced.

The student is probably already familiar with this scheme.

Alpha, ἄλφα, α, A, a as in father

Beta, βῆτα, β, B, b as in boy

Gamma, γάμμα, γ, Γ, g as in go

Delta, δέλτα, δ, Δ, d as in dog
Epsilon, ἐψιλόν, ε, Ε, e as in set
Zeta, ζῆτα, ζ, Ζ, z as in haze
Eta, ἦτα, η, Η, e as in obey
Theta, θῆτα, θ, Θ, th as in think
Iota, ἰότα, ι, Ι, i as in intrigue
Kappa, κάππα, κ, Κ, k as in kite
Lambda, λάμβδα, λ, Λ, l as in leap
Mu, μῦ, μ, Μ, m as in map
Nu,νῦ, ν, Ν, n as in not
Xi, ξῖ, ξ, Ξ, x as in axe
Omicron, ὀμικρόν, ο, Ο, o as in not
Pi, πῖ, π, Π, p as in peek
Rho, ῥῶ, ρ, Ρ, r as in rope
Sigma, σίγμα, σ or ς, Σ, s as in stud
Tau, ταῦ, τ, Τ, t as in take
Upsilon, ὐψιλόν, υ, Υ, y as in German ü
Phi, φῖ, φ, Φ, ph as in photo
Chi, χῖ, χ, Χ, ch as in loch
Psi, ψῖ, ψ, Ψ, ps as in lips
Omega, ὦμέγα, ω, Ω, o as in tone

THE UPPER CASE letters in Greek are called "uncials" and the lower case letters, "minuscules."

Pronunciation of common double vowels or diphthongs is as follows: **αι** (ai as in aisle), **ει** (ei as in weight), **οι** (oi as in foil), **αυ** (au as in how), **ευ** (eu as in feud), **ου** (ou as in coupe), and **υι** (ui as in quite)

2. The Modern Greek Pronunciation scheme, the method we will using in this textbook, is presented below.

Shown, in order, are the English term for the Greek letter, the Greek term for the letter, the Greek lowercase letter, the Greek uppercase letter, and finally, how the letter is pronounced.

Alpha, ἄλφα, , A, a as in father
Beta, βῆτα, β, B, v as in very
Gamma, γάμμα, γ, Γ, g as in go
Delta, δέλτα, δ, Δ, th as in the
Epsilon, ἒψιλόν, ε, E, e as in set
Zeta, ζῆτα, ζ, Z, z as in haze
Eta, ἦτα, η, H, ee as in eel
Theta, θῆτα, θ, Θ, th as in think
Iota, ἰότα, ι, I, ee as in eel
Kappa, κάππα, κ, K, k as in kite
Lambda λάμβδα, λ, Λ, l as in leap
Mu, μῦ, μ, M, m as in map
Nu,νῦ, ν, N, n as in not
Xi, ξῖ, ξ, Ξ, x as in axe
Omicron, ὂμικρόν, ο, O, o as in not
Pi, πῖ, π, Π, p as in peek
Rho, ῥῶ, ρ, P, r as in rope
Sigma, σίγμα, σ or ς, Σ, s as in stud
Tau, ταῦ, τ, T, t as in take
Upsilon, ὒψιλόν, υ, Y, ee as in eel
Phi, φῖ, φ, Φ, ph as in photo
Chi, χῖ, χ, X, ch as in loch
Psi, ψῖ, ψ, Ψ, ps as in lips
Omega, ὢμέγα, ω, Ω, o as in tone

THREE

Μακάριοι οἱ ἄμωμοι ἐν ὁδῷ, οἱ πορευόμενοι ἐν νόμῳ κυρίου.
Blessed are those whose ways are blameless, who walk in the law of the Lord.
Psalm 119:1

3. The forms of the Greek verb "to be"

The present active indicative of the verb, **εἰμί**, ("I am") is the first verb we will study.

Remember that rote memorization of the forms is not our goal. We must only be concerned with recognizing the forms when encountered.

THE STUDENT WILL NOTE that the traditional King James Version of the Bible renders the second person singular as "thou" and the second person plural as "ye."

. . .

SINGULAR

1) (1st person singular) **εἰμί** (I am)
2) (2nd person singular) **εἶ** (you are)
3) (3rd person singular) **ἐστί(ν)** (he/she/ it is)

PLURAL

1) (1st person plural) **ἐσμέν** (we are)
2) (2nd person plural) **ἐστέ** (you are)
3) (3rd person plural) **εἰσί(ν)** (they are)

FROM NOW ON, verb conjugations will list only the number of the particular grammatical person as 1P, 2P, 3P, etc., not the description (1st person singular, etc.).

WHAT IS the grammatical definition of the concept, "present active indicative?" The *present* tense refers to present time; the *active* voice represents the subject as acting (doing something) instead of being acted upon; the *indicative* mood makes an assertion or a statement, in distinction, for example, from a command or a wish. These concepts will be explained in more depth in the next chapter.

Note the forms for the third person singular, **ἐστί(ν)**, and the third person plural, **εἰσί(ν)**. They feature a concept known as the "movable **ν**." The **ν** is often inserted at the end of the verb form when these forms occur at the end of a sentence or when these forms precede a word that begins with a vowel. This latter instance is similar to what happens in English when the article "a" becomes "an" when it precedes a word beginning with a vowel. This is done to make it easier to pronounce the two words in succession.

See the two examples below. Greek word order sometimes differs from English.

ὁ ἀπόστολος ἄνθρωπος ἐστίν." The apostle is a man."

ὁ ἀπόστολός ἐστίν ἀγαθός. "The apostle is good."

The verb **εἰμί** does not take the direct object case, which is called the accusative case. **εἰμί** takes what is called a predicate nominative. The nominative case is called the subject case.

In a sentence such as, "The servant says the word," we can see that it is asserted that the servant does something to the word. This would be an example, then, of the accusative grammatical case, since "the word" is the object of the action expressed by the verb, "says." In this example, "the word" stands in the accusative case, and is the direct object of the verb, "says." However, in the sentence, "The servant is a man," it is not asserted that the servant does anything to the man. "The servant," therefore, stands in the predicate nominative (nominative case), not in the accusative case.

4. Below are some sample verses using the verb **εἰμί**. The occurrences of **εἰμί**, as well as the English equivalents, are bolded and underlined. The symbol ; is the Greek version of our question mark.

1 JOHN 5:5

τίς [δέ] **<u>ἐστιν</u>** ὁ νικῶν τὸν κόσμον εἰ μὴ ὁ
πιστεύων ὅτι Ἰησοῦς **<u>ἐστιν</u>** ὁ υἱὸς τοῦ θεοῦ;
[And] who **<u>is</u>** he who overcomes the world,

but he who believes that Jesus **is** the Son of God?

MARK 2:28

ὥστε κύριός **ἐστιν** ὁ υἱὸς τοῦ ἀνθρώπου καὶ τοῦ σαββάτου.

So that the son of man **is** also lord of the sabbath.

MARK 14:14B

Ὁ διδάσκαλος λέγει, Ποῦ **ἐστιν** τὸ κατάλυμά μου ὅπου τὸ πάσχα μετὰ τῶν μαθητῶν μου φάγω;

The teacher says, "Where **is** the inn, where I might eat the Passover with my disciples?"

MARK 3:11B

Σὺ **εἶ** ὁ υἱὸς τοῦ θεοῦ.

Thou **art** the Son of God.

MARK 5:9

Καὶ ἐπηρώτα αὐτόν, Τί ὄνομά σοι; καὶ λέγει αὐτῷ, Λεγιὼν ὄνομά μοι, ὅτι πολλοί **ἐσμεν**.

And he asked him, "What is your name?" And he says to him, "My name is legion, because we **are** many."

MARK 4:17A

Καὶ οὐκ ἔχουσιν ῥίζαν ἐν ἑαυτοῖς ἀλλὰ πρόσκαιροί **εἰσιν**.
And they have not root(s) in themselves, but **are** temporary.

1 JOHN 4:10

ἐν τούτῳ **ἐστὶν** ἡ ἀγάπη.
In this **is** love.

1 JOHN 4:8B

ὁ θεὸς ἀγάπη **ἐστίν**.
God **is** love.

1 JOHN 4:5

αὐτοὶ ἐκ τοῦ κόσμου **εἰσίν**· διὰ τοῦτο ἐκ τοῦ κόσμου λαλοῦσιν καὶ ὁ κόσμος αὐτῶν ἀκούει.
They **are** of the world. Therefore, they speak of the world, and the world hears them.

JOHN 1:19A

Καὶ αὕτη **ἐστὶν** ἡ μαρτυρία τοῦ Ἰωάννου.
And this **is** the testimony of John.

JOHN 1:19B

Σὺ τίς **εἶ**;
Who **are** you?"

. . .

JOHN 1:34B

οὗτός **ἐστιν** ὁ υἱὸς τοῦ θεοῦ.
This **is** the Son of God.

JOHN 1:49

Ῥαββί, σὺ **εἶ** ὁ υἱὸς τοῦ θεοῦ, σὺ βασιλεὺς **εἶ** τοῦ Ἰσραήλ.
Rabbi, you **are** the Son of God; you **are** King of Israel.

JOHN 14:6

λέγει αὐτῷ [ὁ] Ἰησοῦς, Ἐγώ **εἰμι** ἡ ὁδὸς καὶ ἡ ἀλήθεια καὶ ἡ ζωή· οὐδεὶς ἔρχεται πρὸς τὸν πατέρα εἰ μὴ δι' ἐμοῦ.
Jesus says to him, "I **am** the way, the truth, and the life. No one comes to the Father, except through me.

JOHN 15:1

Ἐγώ **εἰμι** ἡ ἄμπελος ἡ ἀληθινή, καὶ ὁ πατήρ μου ὁ γεωργός **ἐστιν**.
I **am** the true vine, and my Father **is** the farmer.

JOHN 15:3A

ἤδη ὑμεῖς καθαροί **ἐστε**.·
You **are** already pure.

. . .

JOHN 17:16

ἐκ τοῦ κόσμου οὐκ **<u>εἰσὶν</u>** καθὼς ἐγὼ οὐκ **<u>εἰμὶ</u>** ἐκ τοῦ κόσμου.

They **<u>are</u>** not of the world even as I **<u>am</u>** not of the world.

MARK 6:3

οὐχ οὗτός **<u>ἐστιν</u>** ὁ τέκτων, ὁ υἱὸς τῆς Μαρίας καὶ ἀδελφὸς Ἰακώβου καὶ Ἰωσῆτος καὶ Ἰούδα καὶ Σίμωνος; καὶ οὐκ **<u>εἰσὶν</u>** αἱ ἀδελφαὶ αὐτοῦ ὧδε πρὸς ἡμᾶς;

<u>Is</u> not this man the carpenter, the son of Mary and brother of James and Joses and Judah and Simon? And **<u>are</u>** not his sisters here with us?

MARK 8:29B

ἀποκριθεὶς ὁ Πέτρος λέγει αὐτῷ, Σὺ **<u>εἶ</u>** ὁ Χριστός.

And by way of answer Peter says to him, "You **<u>are</u>** the Christ."

1 JOHN 4:6

ἡμεῖς ἐκ τοῦ θεοῦ **<u>ἐσμεν</u>**· ὁ γινώσκων τὸν θεὸν ἀκούει ἡμῶν, ὃς οὐκ **<u>ἔστιν</u>** ἐκ τοῦ θεοῦ οὐκ ἀκούει ἡμῶν. ἐκ τούτου γινώσκομεν τὸ πνεῦμα τῆς ἀληθείας καὶ τὸ πνεῦμα τῆς πλάνης.

We **<u>are</u>** of God. He who knows God listens to us. He who **<u>is</u>** not of God doesn't listen to us.

By this we know the spirit of truth, and the
spirit of error.

5. **Recognition Drill**

Match the English with the Greek translation. Answers are provided below.

a) He is a man./b) Thou art good./c) He is an apostle./ d) I am a man./e) The apostle is a man./f) The man is good./g) I am an apostle./h) Thou art a man./i) I am good./j) He is good.

_____ 1. ὁ ἀπόστολος ἄνθρωπος ἐστίν.
_____ 2. ὁ ἀπόστολός ἐστὶν ἀγαθός.
_____ 3. εἰμὶ ἀπόστολός.
_____ 4. εἶ ἄνθρωπος.
_____ 5. ἐστὶν ἀγαθός.
_____ 6. ἐστὶν ἀπόστολος.
_____ 7. εἶ ἀγαθός.
_____ 8. εἰμὶ ἀγαθός.
_____ 9. εἰμὶ ἄνθρωπος.
_____ 10. ἐστὶν ἄνθρωπος.

Answers: 1e, 2f, 3g, 4h, 5j, 6c, 7b, 8i, 9d, 10a

FOUR

Γινώσκει κύριος ὁδὸν δικαίων.
The Lord knows the way of the righteous.
Psalm 1:6

The Present Active Indicative Tense

6. Vocabulary, verbs

ἄγω, I lead, I am leading, I do lead; this verb can also mean "I bring."

ἀκούω, I hear, am hearing, do hear ; can also mean "I listen." (acoustic)

βλέπω, I see, am seeing, do see

γινώσκω, I know, am knowing, do know (The second syllable, **νώ**, sounds like "know.")

γράφω, I write, am writing, do write (graph)

διδάσκω, I teach, am teaching, do teach (didactic)

ἐγείρω, I raise up, am raising up, do raise up
ἔχω, I have, am having, do have
λαμβάνω, I take, am taking, do take
λέγω, I say, am saying, do say; I speak
λύω, I loose, am loosing, do loose
πέμπω, I send, am sending, do send
φέρω, I bear, I bring, carry, etc. (Christopher means "Christ bearer")

7. Vocabulary, nouns

Take special note of the cognates in parentheses below.

ἄγγελός, οῦ, ὁ, angel, messenger (angel)

ἀδελφός, οῦ, ὁ, brother (Philadelphia) The **οῦ** after the noun indicates the genitive (possessive) case form and the **ὁ** after that signifies the masculine definite article, that is, "the." More on these matters later.

ἄνθρωπος, ου, ὁ, man, human being (anthropology)
ἀπόστολος, ου, ὁ, apostle
δοῦλος, ου, ὁ, servant, slave

δῶρον, ου, τό, gift (Dorothy means "gift of the goddess") **τό** is the neuter Greek word for "the."

καρπός, οῦ, ὁ, fruit
λόγος, ου, ὁ, word (logical)

8. **Recognition Tip**

In the above vocabulary list, and in all subsequent vocabulary listings, the student will notice the words in parentheses after a few of the terms. These words are known as "cognates." A cognate is a term in two or more languages, which means exactly (or about) the same thing in all those languages.

Cognates represent an effective aid to recognizing and learning Greek vocabulary.

9. All languages have tense. Tense is that quality of a verb which indicates action. There are two elements of action: *time* of action and *kind* of action. The *time* of the action can be present, past or future. The *kind* of action can be linear or puncticular. Linear action is progressive or continuous action as in "She is reading." Puncticular action is restricted to a particular moment in time or a singular occurrence, such as "He wrote."

AS MENTIONED ABOVE, the present tense (in the indicative, which means a simple statement of fact), can indicate progressive or continuous action at the present time. Depending on the context, the more correct translation of the verb forms listed above could be, for example, **ἄγω**, I am leading, but translations of "I lead" or "I do lead," could also be acceptable.

IN THE ACTIVE voice the subject of the sentence does the acting as opposed to being acted upon.

IN THE INDICATIVE mood the subject of the sentence employs the verb to make an assertion as opposed to expressing a command or a wish. For example, "They are leading the sheep" indicates that the action is really taking place at the present time. "Lead my sheep" or "I hope you lead my sheep" indicate that the action of the verb is not really taking place at this time, but is potential – that is, it is possible for the action to be taking place at this particular moment.

PERSON AS RELATING to verbs indicates whether the subject of the sentence is speaking (first person, "I am listening"), is being addressed or spoken to (second person, "You are listening"), or is being spoken of (third person, "He [she or it] is listening").

NUMBER AS RELATING to verbs indicates whether the subject is singular ("I am listening") or plural ("We are listening").

10. Our sample Greek verb is **λύω** which means "I loose, I am loosing, I do loose." This verb can also mean "I destroy, I am destroying, I do destroy."

Below is the conjugation of the present active indicative of this verb. As mentioned above, the student will see that each verb consists of Singular and Plural forms (Number) and that each of these is divided into "Persons," such as 1p, 2P, etc.

The student will recall that the first person is the person speaking, the second person is the person spoken to, and the third person is the person spoken of.

SINGULAR

1P **λύω** (I loose)
2P **λύεις** (you loose)
3P **λύει** (he/she/it looses)

. . .

PLURAL

1P **λύομεν** (we loose)
2P **λύετε** (you loose)
3P **λύουσι (ν)** (they loose)

THE SECOND PERSON FORMS, **λύεις** (you loose) and **λύετε** (you loose) are actually the King James Version forms "thou art loosing" (**λύεις**) and "ye are loosing" (**λύετε**), respectively. As mentioned before, notice that in the third person plural form the Greek letter "**ν**" is sometimes added when this form comes at the end of a sentence or when it precedes a word that begins with a vowel.

NOTE that these forms combine the concept of subject (in this case, pronoun) and verb in one word. Therefore, no pronoun is required to translate the English phrase "we loose" (**λύομεν**) into Greek. In this example, "we" is adequately represented by the verb ending "**ομεν**." To be sure, a separate pronoun can be used with these verb forms, but this use will be reserved for future consideration.

NOTE ALSO that the third person forms of verbs can be combined with a noun: **ὁ ἄνθρωπος λύει** means "the man looses," whereas **λύει** standing by itself means "he (she or it) looses" (is loosing).

THE "STEM" is that portion of the verb which does not change throughout the conjugation or parsing of the verb. The present tense stem of the verb **λύω** is **λύ-** and the present tense stem of the verb **λέγω** (I say, am saying, do say) is **λέγ**-. To determine the stem of a verb, simply remove the final **ω** from the verb form given in the vocabulary lists. The present active indicative forms of any of the above-listed verbs can be obtained by taking the stem of the verb and adding the endings seen after the stem: **ω**, **εις**, etc. These are the endings of the primary tenses. (The primary tenses are the present, the future and the perfect tenses. The secondary tenses are the imperfect, the aorist and the pluperfect tenses. We will deal more with these concepts in future lessons.)

NOTICE that in the present tense forms, the Greek language does not differentiate between "I loose" and "I am loosing." "I loose" simply means that the action of the verb is occurring in present time. "I am loosing" calls to mind the continuance of the action of the verb.

11. Recognition Tips

Take note of the following clues, which apply to verbs of the present tense, such as **λύω**, the forms of which were given in section 10, above:

The first person singular ends in **ω**, which means that **ω** is essentially equivalent to the pronoun "I."

The second person singular ends in **εις**, which means that **εις** is essentially equivalent to the pronoun "you" or "thou."

The third person singular ends in **ει**, which means that **ει** is essentially equivalent to the pronoun "he," "she," or "it."

The first person plural ends in **μεν**, which means that **μεν** is essentially equivalent to the pronoun "we." The **ο** added between the stem of the verb (**λύ-**) and the ending **μεν**, is called a variable vowel, added to make it easier to pronounce the word.

The second person plural ends in **τε** , which means that **τε** is essentially equivalent to the pronoun "you" or "ye." The **ε** added between the stem of the verb (**λύ-**) and the ending **τε**, is a variable vowel.

The third person plural ends in **ουσι (ν)**, which means that **ουσι (ν)** is essentially equivalent to the pronoun "they."

Remember these tips regarding verb endings as you encounter other present tense verbs. Also keep in mind that the verbs listed at the beginning of this chapter are all conjugated like **λύω**.

12. **Recognition Drill. Present Active Indicative Practice with Answers**

Working down the page, cover the Greek term with a 3 X 5 card and translate each form into English. The English answers are directly under the Greek. Keep in mind the clues regarding verb endings mentioned above in the **Recognition Tips** in section 11.

ἄγω
I lead.
ἄγεις
You lead.
ἀκούομεν
We hear/listen.
βλέπετε
Ye see.
γινώσκουσι (ν)
They know.
διδάσκει
He/she/it teaches.

γράφεις
You write.
ἔχω
I have.

USING THE SAME METHOD, translate the following English sentences into Greek. Do not be concerned that the accent marks move around within a Greek word. This will be explained later.

I send.
πέμπω
He/she/it carries.
φέρει
We say.
λέγομεν
Ye take.
λαμβάνετε
They have.
ἔχουσι(ν)
Thou writest.
γράφεις
I take.
λαμβάνω
We take.
λαμβάνομεν

FIVE

Θύσατε θυσίαν δικαιοσύνης καὶ ἐλπίσατε ἐπὶ κύριον.
Offer a sacrifice of righteousness, and hope in the Lord.
Psalm 4:5

Definition of Voices

13. In the active voice the subject of the verb is the doer of the action of thar verb, i.e., the action of the verb moves outward from the subject. This could be described diagrammatically as follows:

VERB
SUBJECT --------------------------->

. . .

FOR EXAMPLE: "The woman sees the temple."

14. In the passive voice the subject of the sentence is the thing upon which the action of the verb is carried out.

VERB
SUBJECT <--------------------------

FOR EXAMPLE:
"They shall be comforted."
"The word was spoken by the prophet."

15. The middle voice, as the name suggests, lies between the active and the passive voice in function. There are two basic uses which meld or shade into one another:

1) IT IS COMMONLY USED for intransitive verbs or where the action of the verb does not carry over to an object, but solely affects the subject. Often the subject of a middle-voice verb is acting for himself or in his own interests as in Mark 6:7a: "And He called the twelve to Himself ..."

2) THE SECOND use is closely related to the first use. In this case the middle voice can have a reflective sense, in other words, the action was done to or for the subject.

Reflective use: Matt. 27:5: "Judas went and hanged himself." (Middle Voice)

Note, then, that most commonly the middle voice is rendered in English by the active voice, with the added pronoun (himself, for himself, etc.), and at times with a construction that indicates that the subject had the action carried out upon himself.

SIX

Κύριε ὁδήγησόν με ἐν τῇ δικαιοσύνῃ σου.
Lord, lead me in your righteousness.
Psalm 5:8

16. **Present Active Indicative Verse Examples**

PRESENT ACTIVE INDICATIVE GREEK VERBS, as well as the English equivalents, are bolded and underlined.

Concentrate on **<u>recognizing</u>** the endings on the underlined Greek verbs for clues as to the pronoun associated with that ending.

1 JOHN 5:12

ὁ ἔχων τὸν υἱὸν **<u>ἔχει</u>** τὴν ζωήν· ὁ μὴ ἔχων
τὸν υἱὸν τοῦ θεοῦ τὴν ζωὴν οὐκ **<u>ἔχει</u>**.

He who **has** the Son has the life. He who doesn't have God's Son doesn't **have** the life.

MARK 1:12

Καὶ εὐθὺς τὸ πνεῦμα αὐτὸν **ἐκβάλλει** εἰς τὴν ἔρημον.

And immediately the Spirit **drives** him **out** into the wilderness.

MARK 3:33

Καὶ ἀποκριθεὶς αὐτοῖς **λέγει**, Τίς ἐστιν ἡ μήτηρ μου καὶ οἱ ἀδελφοί [μου];

And answering them, he **says**, "who is my mother and my brothers?"

MARK 4:14

ὁ σπείρων τὸν λόγον **σπείρει**.

The one sowing the word, **sows**.

MARK 4:13

Καὶ **λέγει** αὐτοῖς, Οὐκ **οἴδατε** τὴν παραβολὴν ταύτην;

And he **says** to them, "don't you **know** this parable?"

MARK 3:13A

Καὶ **ἀναβαίνει** εἰς τὸ ὄρος.

And he **goes up** into the mountain.

. . .

MARK 11:22A

Καὶ ἀποκριθεὶς ὁ Ἰησοῦς **λέγει** αὐτοῖς, **Ἔχετε** πίστιν θεοῦ.

And by way of answer, Jesus says to them, "Have faith in God."

MARK 14:13A

Καὶ **ἀποστέλλει** δύο τῶν μαθητῶν αὐτοῦ καὶ **λέγει** αὐτοῖς, **Ὑπάγετε** εἰς τὴν πόλιν.

And he **sends** out two of his disciples and **says** to them, "**Go** into the city."

MARK 3:30

πνεῦμα ἀκάθαρτον **ἔχει**.

He **has** an unclean spirit. (Did you **recognize** that the **ει** on **ἔχει** signaled the third person singular pronoun, "he"?)

JOHN 10:15

καθὼς **γινώσκει** με ὁ πατὴρ κἀγὼ **γινώσκω** τὸν πατέρα.

Just as the Father **knows** me, I also **know** the Father. (Did you **recognize** that the **ει** on the end of **γινώσκει** signaled the third person singular pronoun, "he"?) In addition, the **ω** at the end of **γινώσκω** ("I know") signals the first person singular pronoun, "I."

. . .

JOHN 4:36

καὶ ὁ θερίζων μισθὸν **λαμβάνει** καὶ **συνάγει** καρπὸν εἰς ζωὴν αἰώνιον.
The one who reaps, **receives** wages and **gathers** fruit for eternal life.

JOHN 6:20

ὁ δὲ **λέγει** αὐτοῖς, ἐγώ **εἰμι**· μὴ φοβεῖσθε.
And he **says** to them, "It **is** I, don't be afraid."

1 JOHN 2:1

Τεκνία μου, ταῦτα **γράφω** ὑμῖν ἵνα μὴ ἁμάρτητε· καὶ ἐάν τις ἁμάρτῃ, παράκλητον **ἔχομεν** πρὸς τὸν πατέρα, Ἰησοῦν Χριστὸν δίκαιον·
My little children, I **write** these things to you so that you might not sin. And if anyone should sin, we **have** an advocate with the Father, Jesus Christ the righteous.

1 JOHN 4:18

φόβος οὐκ **ἔστιν** ἐν τῇ ἀγάπῃ, ἀλλ' ἡ τελεία ἀγάπη ἔξω **βάλλει** τὸν φόβον, ὅτι ὁ φόβος κόλασιν **ἔχει**.
There **is** no fear in love, but [the] perfect love **casts** out fear, because fear **involves** [has] torment.

SEVEN

εἰσήκουσεν κύριος τῆς δεήσεώς μου.
The Lord has heard my cry for mercy.
Psalm 6:9a

The Second Declension of Greek Nouns

17. New Testament Greek has three noun declensions. The second declension is normally dealt with before the first and third declensions because it is easier to learn and encompasses a larger number of nouns. The third declension is not covered in this book. Listed below are some high-frequency second declension nouns as well as some common conjunctions. You have seen some of these nouns earlier in the book. Learn to **recognize** them all. Note also that each noun is followed by a Greek word for "the," also known as the definite article. **ὁ** is the

masculine word for "the," **ἡ** is the feminine and, **τό** is the neuter.

Note, also, as mentioned before, that many of the words in the vocabulary list have cognates in English, which will aid in **recognition.**

ἄγγελος, ου, ὁ angel; messenger

ἀδελφός, οῦ, ὁ brother (Philadelphia)

ἄνθρωπος, ου, ὁ man; person (anthropology)

ἄρτος, ου, ὁ bread

δέ but; and (conjunction)

δοῦλος, ου, ὁ servant; slave

ἔθνος, ους, τό nations; gentiles (ethnic)

ἔργον, ου, τό work; deed

θάνατος, ου, ὁ death (thanatology is the scientific study of death)

θεός, οῦ, ὁ God (theology)

καί, and, also (conjunction)

καιρός, οῦ, ὁ time; period

κάρπος, οῦ, ὁ fruit

κόσμος, ου, ὁ world (cosmos)

κύριος, ου, ὁ Lord

λίθος, ου, ὁ stone (lithosphere)

λαός, οῦ, ὁ people

λόγος, ου, ὁ word (logic)

νόμος, ου, ὁ law (nomology)

ὁδός, οῦ, ἡ way (odometer [measures the "way"])

οἶκος, ου, ὁ house

οὐρανός, οῦ, ὁ heaven (Uranus)

ὀφθαλμός, οῦ, ὁ eye (ophthalmologist)

ὄχλος, ου, ὁ crowd

Παῦλος, ου, ὁ Paul

Πέτρος, ου, ὁ Peter

Πιλᾶτος, ου, ὁ Pilate

τέκνον, ου, τό child (technology)
τόπος, ου, ὁ place (topography)
Φαρισαῖος, ου, ὁ Pharisee
Χριστός, οῦ, ὁ Anointed; Christ
υἱός, οῦ, ὁ son

THE NOUNS in the above vocabulary are a mix of masculine, feminine and neuter nouns. Many of the masculine nouns refer to men, but some refer to things, which makes those things grammatically masculine. There are also many feminine nouns in Greek which refer to women, as well as many other feminine nouns which refer to things and have nothing to do with physical gender.

18. **Recognition Tip**

THE GENDER of each individual noun must be learned by observation, but almost every noun belonging to the second declension, which ends in -**ος**, is a masculine noun, which means that the definite article (the word for "the") which belongs to second declension nouns ending in -**ος**, is **ὁ**. Furthermore, all nouns of the second declension ending in -**ον** are neuter, which means that the definite article (the word for "the") which belongs to second declension nouns ending in -**ον** is **τό**.

IN THE VOCABULARY ABOVE, the definite article is placed after the noun. Below, in section 20, when the declension of **ἄνθρωπος** (and later, other nouns) is given, the articles are placed before the nouns to replicate the manner in which the student will encounter the articles and nouns in the text of the New Testament.

19. In addition to gender, all Greek nouns have number and case.

THERE ARE TWO NUMBERS: singular and plural. In other words, a noun can be singular (woman) or plural (women). Verbs agree with their subject in number, whether the subject is a noun or a pronoun. In other words, a singular noun will be connected to a singular form of a verb, and a plural noun to a plural verb form.

THE CONCEPT OF CASE, of which there are five in Greek, will be explained more clearly after we consider the noun forms, below.

20. The declension of **ἄνθρωπος** "man/human being," is as follows:

SINGULAR

Nominative **ὁ ἄνθρωπος**, the man
Genitive **τοῦ ἀνθρώπου**, of the man
Dative **τῷ ἀνθρώπῳ**, to/for the man
Accusative **τὸν ἄνθρωπον**, the man
Vocative **ἄνθρωπε**, man

. . .

PLURAL

Nominative **ὁι ἄνθρωποι**, the men
Genitive **τῶν ἀνθρώπων**, of the men
Dative **τοῖς ἀνθρώποις**, to/for the men
Accusative **τοὺς ἀνθρώπους**, the men
Vocative **ἄνθρωποι**, men

21. **Noun stems**

THE STEM of the noun is that part of the noun which does not change when the various noun endings are added. The stem of **ἄνθρωπος** is **ἄνθρωπο-**. For this reason the second declension is sometimes called the **ο**-declension. However, the final **ο** becomes so much disguised when the noun endings are combined with it, that it is better to consider **ἄνθρωπ-** as the stem of this noun and **-ος**, **-ου**, etc., as the endings. In any case **ο**, and its long form **ω** is the characteristic vowel in the last syllable of nouns of the second declension.

22. **Explanation of Cases**

IN GREEK, as in English, putting a noun in a particular case defines the meaning and usage of that noun. Greek uses five basic cases. English uses the same cases, but they are more difficult to recognize. For the sake of simplicity, we will not discuss the cases in the exact order that they appear in the chart above which dealt with **ἄνθρωπος**.

. . .

23. **Nominative Case**

THE SUBJECT of a sentence is in the nominative case; thus the nominative case is also called the subject case. In this grammatical case the subject is usually doing the "acting" by means of the verb connected with it. Consider this English sentence: "The apostle knows." Here, the subject is "the apostle" and the apostle is doing something. Now, let us look at a Greek sentence in which the nominative form (see the declension above) serves as the subject of the sentence: **ὁ ἄνθρωπος βλέπει** ("The man sees."). This an example of the nominative singular. The subject is singular (one man) and the man is the subject of the sentence, and thus, grammatically we say that this sentence is in the nominative case. Let us consider a sentence in the subject (nominative) case in which there is a plural subject, i.e., more than one man: **οἱ ἄνθρωποι βλέπουσι** ("The men see.") In this case there is more than one man and these men are the subject of the sentence.

24. **Recognition Tip**

IN THE TWO sentences cited above, remember that **βλέπει** is third person singular because of the verb ending **ει** and can mean "he sees," "she sees," or "it sees," when stated by itself. However, when **βλέπει** is combined with a singular noun, such as **ὁ ἄνθρωπος** to form **ὁ ἄνθρωπος βλέπει**, it is translated as, "the man sees."

25. **Genitive Case**

. . .

THE GENITIVE CASE is often the case of possession. In other words this case expresses the fact that someone or something owns something. Therefore **ὁ λόγος τῶν ἀνθρώπων** means, "The word of the men." In this case the subject of the sentence is **ὁ λόγος** (the word). As we progress through the course, we will see that the genitive case has other important uses.

26. **Accusative Case**

THE OBJECT of a transitive verb is placed in the accusative case. A transitive verb is one that exerts its force on a noun or pronoun.

Let's see how this works in English. Consider the sentence, "The apostle sees the man." In this case the transitive verb "sees" is acting upon the man. The object (sometimes called the direct object) of the verb "sees" is "the man." How can one determine which noun or pronoun in a sentence is the direct object? In this sample sentence, simply look at the verb, in this case, "sees," and ask the question, "What or whom does the apostle see?" In this case the apostle sees the man. Therefore "the man" is the direct object of the verb "sees."

IN GREEK, the form of the definite article and of the noun that together comprise the direct object undergo a change if they are masculine or feminine, whether they are singular or plural. This change in the form of the definite article and noun allows the reader to easily identify which word or words are the direct object, that is, the object of the transitive verb. Here is our sample sentence, "The angel sees the house," in Greek:

ὁ ἄγγελος βλέπει τὸν οἶκον.

NOTICE that we have transferred the exact form of the Greek word for "the angel" from the vocabulary list, and that "angel" is the subject form of this noun in the singular. The accusative form for house ends in **ν** in the singular, just as **ἄνθρωπον** in the chart above ends in **ν**. We ask the question, "What does the angel see?" The answer is "the house." Therefore "the house" in both Greek and English is the direct object, the object of the transitive verb "sees." This means that "the house" is in the accusative case and must be rendered as **τὸν οἶκον**.

AS WITH THE GENITIVE CASE, the accusative case has other uses, which will be discussed later.

27. **Dative Case**

ONE OF THE functions of this case lies in its use as the case of the indirect object. Notice from the declension chart that the word "to" or "for" is often connected to nouns in the dative. The noun or pronoun that is connected to "to" or "for" is the indirect object. Therefore, in this instance, the dative case answers the question "to what/whom?" or "for what/whom?" Here is an example in English:

"I say the word to the men."

In this sentence "I" is the subject of the sentence. "I say" is in the nominative, or subject, case; "the word" is the direct object of the verb "say" and is in the accusative case. (What do I say? I say the word. Therefore "the word" is the direct object.); "to the men" is the indirect object of the verb "say" and is in the dative case. (**To** whom do I say the word? I say the word **to** the men.) Therefore "to the men" is the indirect object. Here is the same sentence in Greek:

λέγω τὸν λόγον τοῖς ἀνθρώποις.

AS IN THE case of the genitive and accusative cases, the dative case has other functions, as we will learn later.

28. **Vocative Case**

THE VOCATIVE CASE is fairly rare in Greek and is the case of direct address. Thus **ἄνθρωποι, βλέπομεν** means, "men, we see."

29. Please note the difference between a definite article, which in English is designated by the word "the," as in "the women," and the indefinite article. In the case of the English indefinite article, we employ the word "a," as in "a woman" or we simply say "woman" without the indefinite article "a." There is no indefinite article in Greek so **ἄνθρωπος** can mean either "human" (man) or "a human" (a man). In a few cases **ἄνθρωπος** standing by itself can mean "the man," but the normal meaning for the indefinite article is "a man" or simply "man." Plural nouns in Greek and in English have no indefinite article, so the plural noun **ἄνθρωποι** means "men," but not "the men."

. . .

30. **Recognition Tips**

NOW THAT WE have explained the various cases, let us consider some specific examples.

ἄνθρωπος λέγει: "A man speaks." Nominative singular.

ἀπόστολος βλέπει ἄνθρωπον: "An apostle sees a man." In this case, **ἀπόστολος** is in the nominative singular (the subject of the sentence) and **ἄνθρωπον** is in the accusative singular since it is the direct object.

λέγω τὸν λόγον τοῖς ἀποστόλοις. "I speak (say) the word to the apostles." Dative plural.

λέγει τοὺς λόγους τῶν ἀποστόλων. "She speaks the words of the apostles." Genitive plural.

λέγεις λόγον τῷ ἀνθρώπῳ. "Thou speakest a word to the man." Dative singular.

IMPORTANT FACT: The dative singular in all three genders and in all Greek noun declensions is easily recognizable because the forms end in **ι**. Consider this sentence: **λέγομεν λόγον τῷ ἀνθρώπῳ.**

The sentence means, "We speak a word to the man." The dative portion of the sentence is **τῷ ἀνθρώπῳ**, meaning, "to the man." Sometimes the final **ι** is the letter **ι** in a word, and sometimes, as in "**τῷ ἀνθρώπῳ**" the ending **ι** appears as what is called the iota-subscript, that is, it appears under the final vowel of the definite article and the noun in question. Therefore: "**τῷ ἀνθρώπῳ**."

. . .

31. The chart listed below is the same chart presented in section 20, above, but with one important difference: the definite articles have been omitted. This changes the meanings of the English nouns from "the man," for instance, to "a man." Thus, the word "the" is called the definite article, because we are referring to a specific man, whereas the word "a" is the indefinite article, because it refers to no man in particular.

SINGULAR

Nominative **ἄνθρωπος**, a man
Genitive **ἀνθρώπου**, of a man
Dative **ἀνθρώπῳ**, to/for a man
Accusative **ἄνθρωπον**, the man
Vocative **ἄνθρωπε**, man

PLURAL

Nominative **ἄνθρωποι**, men
Genitive **ἀνθρώπων**, of men
Dative **ἀνθρώποις**, to/for men
Accusative **ἀνθρώπους**, men
Vocative **ἄνθρωποι**, men

32. The declension of the neuter second declension noun, **δῶρον, τό**, the gift, is presented below.

Note that in most cases, as is the case in masculine nouns, the endings on the nouns are identical to the definite articles. This fact greatly aids the student in recognizing the various forms of the noun.

. . .

SINGULAR

Nominative **τὸ δῶρον**, the gift
Genitive **τοῦ δώρου**, of the gift
Dative **τῷ δώρῳ**, to/for the gift
Accusative **τὸ δῶρον**, the gift
Vocative **δῶρον**, gift

PLURAL

Nominative **τὰ δῶρα**, the gifts
Genitive **τῶν δώρων**, of the gifts
Dative **τοῖς δώροις**, to/for the gifts
Accusative **τὰ δῶρα**, the gifts
Vocative **δῶρα**, gifts

IN ALL NEUTER NOUNS, of all declensions, the vocative and accusative of both numbers (singular and plural) are like the nominative forms. In addition, the nominative, vocative and accusative plural forms always end in short **α**. Do not be concerned about long and short vowels. They play a role in where the accent mark is placed on a word, but almost no modern New Testament Greek textbooks deal with Greek rules of accent.

33. Accent marks seem to do strange things. In the above chart the words for "the" precede the nouns, and the accent marks on **τό** and **τά** have leaned over to the left away from the noun they precede to become **τὸ** and **τὰ**. This is normal. Accent marks going from upper right to lower left over a vowel are called acute accent marks. When the acute accent is on the last

syllable of a word, as in the case of **τό** and **τά**, they normally lean over as shown above if they precede another word.

34. Recognition Drill

Match the English with the Greek translation. Answers are provided below.

a) I say the word to the apostles./b) We speak a word to the man./c) I see the gifts of the apostles./d) She has the bread of the brothers./e) You hear the word of the servant./f) You speak a word to the man./g) An apostle sees a man./h) I know the house of the messenger./i) A man speaks./j) He speaks the words of the brethren.

_____ **1. ἄνθρωπος λέγει.**
_____ **2. ἀπόστολος βλέπει ἄνθρωπον.**
_____ **3. λέγω τὸν λόγον τοις ἀποστόλοις.**
_____ **4. λέγει τοὺς λόγους τῶν ἀδέλφων.**
_____ **5. λέγεις λόγον τῷ ἀνθρώπῳ.**
_____ **6. λέγομεν λόγον τῷ ἀνθρώπῳ.**
_____ **7. βλέπω τὰ δῶρα τῶν ἀποστόλων.**
_____ **8. ἔχει τὸν ἄρτον τῶν ἀδέλφων.**
_____ **9. γινώσκω τὸν οἶκον ἀγγέλου.**
_____ **10. ἀκούεις τὸν λόγον δούλου.**

Answers: 1i, 2g, 3a, 4j, 5f, 6b, 7c, 8d, 9h, 10e

EIGHT

Ἐν ἀρχῇ ἦν ὁ Λόγος, καὶ ὁ Λόγος ἦν πρὸς τὸν Θεόν,
καὶ Θεὸς ἦν ὁ Λόγος.
In the beginning was the Word, and the Word was with God
and the Word was God.
John 1:1

35. **Second Declension Exercises**

Cover the English answers below the Greek text and translate the following Greek sentences into English.

1) **ἄνθρωπος λέγει λόγον.**
A man says a word.

2) **λόγον λέγει ἄνθρωπος.**
A man says a word.

3) **ἀπόστολος βλέπει τὰ δῶρα.**
An apostle sees the gifts.

4) **ὁ ἄνθρωπος βλέπει τὸν οἶκον.**
The man sees the house.

5) **οἱ ἄνθρωποι βλέπουσι τὸ δῶρον.**
The men see the gift.

NOW COVER THE GREEK TEXT, above, and translate the English sentences into Greek.

36. Recognition Tip

Did you notice that in sentence 2, above, the direct object is the first word in the sentence? Simply stated, **λόγον** is the direct object form of **λόγος** and **ἄνθρωπος** is the subject of the sentence.

37. Additional Present Active Indicative Verses

Applicable Greek verbs, as well as the English equivalents, are bolded and underlined.

Some verses have been repeated from previous exercises.

MARK 3:30

πνεῦμα ἀκάθαρτον **<u>ἔχει</u>**.
He **<u>has</u>** an unclean spirit.

JOHN 1:29

Τῇ ἐπαύριον **<u>βλέπει</u>** ὁ Ἰωάννης τὸν Ἰησοῦν ἐρχόμενον πρὸς αὐτόν.
The next day (the) John **<u>sees</u>** (the) Jesus coming toward him.

. . .

JOHN 10:15

καθὼς **γινώσκει** με ὁ πατὴρ κἀγὼ **γινώσκω** τὸν πατέρα.

Just as the Father **knows** me, I also **know** the Father.

JOHN 4:36

καὶ ὁ θερίζων μισθὸν **λαμβάνει** καὶ **συνάγει** καρπὸν εἰς ζωὴν αἰώνιον.

And the one who reaps **receives** wages and **gathers** fruit for eternal life.

JOHN 6:20

ὁ δὲ **λέγει** αὐτοῖς· ἐγώ **εἰμι**· μὴ φοβεῖσθε.

And He **says** to them, "It **is** I, do not be afraid."

1 JOHN 2:1

Τεκνία μου, ταῦτα **γράφω** ὑμῖν ἵνα μὴ ἁμάρτητε· καὶ ἐάν τις ἁμάρτῃ, παράκλητον **ἔχομεν** πρὸς τὸν πατέρα, Ἰησοῦν Χριστὸν δίκαιον·

My little children, I **write** these things to you so that you might not sin. And if anyone should sin, we **have** an advocate with the Father, Jesus Christ the righteous.

38. **Recognition Drill**

Match the English translation with the Greek text. Answers are provided below.

a) You write the words of the apostles./b) Thou hast the people of God./c) Ye send the gifts of the apostles./d) They know the words of the apostles./e) You lead the servants of God./f) She teaches the law of the Lord./g) They carry the son of the brother./h) You know the house of the messenger./i) We take the fruit of the crowd./j) I raise up the word of the apostles.

_____ **1. ἄγεις τοὺς δούλους θεοῦ.**

_____ **2. γινώσκουσιν τοὺς λόγους τῶν ἀποστόλων.**

_____ **3. Διδάσκει τὸν νόμον τοῦ κυρίου.**

_____ **4. γράφεις τοὺς λόγους τῶν ἀποστόλων.**

_____ **5. ἐγείρω τὸν λόγον τῶν ἀποστόλων.**

_____ **6. λαμβάνομεν τὸν κάρπον τοῦ ὄχλου.**

_____ **7. πέμπετε τὰ δῶρα τῶν ἀποστόλων.**

_____ **8. φέρουσιν τὸν υἱὸν τοῦ ἀδέλφου.**

_____ **9. γινώσκεις τὸν οἶκον ἀγγέλου.**

_____ **10. ἔχεις τὸν λαόν θεοῦ.**

Answers: 1e, 2d, 3f, 4a, 5j, 6i, 7c, 8g, 9h, 10b

NINE

Κύριε ὁ θεός μου ἐπὶ σοὶ ἤλπισα, σῶσόν με.
Lord, my God, upon you I hoped, save me.
Psalm 7:1

Greek Word Order

39. Recognition Tips

The normal order of the Greek sentence is identical to English: subject, verb, object. However, Greek sometimes varies the word order for the purpose of emphasis or euphony (gracefulness or style) much more readily than English. Therefore, the Greek sentence, "the angel sees the house," is normally:

ὁ ἄγγελος βλέπει τὸν οἶκον.

Nonetheless, the same Greek statement can just as easily be rendered as:

βλέπει ὁ ἄγγελος τὸν οἶκον.

Or as:

τὸν οἶκον βλέπει ὁ ἄγγελος.

All three Greek sentences mean, "the angel sees the house." In some cases, therefore, the English translation of a Greek sentence must be rendered by observing the endings of the words to determine their function in the sentence, not by observing the order of the words. Keep in mind, as illustrated above, that a direct object or an indirect object can begin a Greek sentence.

In all three of the above examples, **ὁ ἄγγελος** can only be the subject of the sentence and **τὸν οἶκον** can only be the masculine singular direct object. Notice that the sign of the masculine and feminine accusative singular is the letter **ν**, as in **τὸν οἶκον**, above.)

TEN

ἐξομολογήσομαι κυρίῳ κατὰ τὴν δικαιοσύνην αὐτοῦ.
I will give thanks to the Lord according to his righteousness.
Psalm 7:17a

40. The First Declension

VOCABULARY

ἀγάπη, ἡ, love; kiss of peace

ἀλήθεια, ἡ, truth, truthfulness, faithfulness (Alice)

ἁμαρτία, ἡ, sin (Hamartiology is the study of the nature of sin.)

βασιλεία, ἡ, kingdom (basilica)

γλῶσσα, ἡ, language, tongue (the physical organ); bar, ingot (glossary)

γραφή, ἡ, written document, holy scripture (graphic)

διδαχή, ἡ, teaching (didactic)

δόξα, **ἡ,** opinion; glory (doxology)
εἰρήνη, **ἡ,** peace (irenic, Irene)
ἐκκλησία, **ἡ,** church congregation, assembly, church (ecclesiastic)
ἐντολή, **ἡ,** commandment
ἡμέρα, **ἡ,** day, daylight hours, lifetime, time period (ephemeral means short-lived)
καρδία, **ἡ,** heart (cardiac)
μαθητής, **ὁ,** disciple, student
παραβολή, **ἡ,** proverb, poem, parable
προφήτης, **ὁ,** prophet
συναγωγή, **ἡ**, collection, gathering, synagogue
φωνή, **ἡ,** voice, sound, speech (phone)
ὥρα, **ἡ,** hour (The Greek word sounds like "hour.")

41. **Recognition Tip**

Remember that you only have to recognize the above forms, not memorize them.

In the above vocabulary, **ἡ**, the definite article (a word for "the"), is seen again. This small word was briefly introduced during the discussion of the second declension. **ἡ** is the feminine article. Most of the Greek nouns that designate women (wife, daughter, etc.) are feminine in Greek and carry this article. But notice that many nouns in the above list are feminine and yet have nothing to do with physical gender. All first declension nouns that end in **η** or **α** in the nominative singular are feminine.

First declension masculine nouns end in **ας** or **ης**. Examples of this concept from the above list are **μαθητής**, **ὁ,** "disciple, student" and **προφήτης**, **ὁ,** "prophet."

μαθητής is an interesting noun because we translate it into English as "disciple," but it literally means "learner."

. . .

42. Note below the declension of **ἡ ὥρα** (the hour).

SINGULAR

Nominative **ἡ ὥρα**, the hour
Genitive **τῆς ὥρας**, of the hour
Dative **τῇ ὥρᾳ**, to/for the hour
Accusative **τὴν ὥραν**, the hour
Vocative **ὥρα**, hour

PLURAL

Nominative **αἱ ὥραι**, the hours
Genitive **τῶν ὡρῶν**, of the hours
Dative **ταῖς ὥραις**, to/for the hours
Accusative **τὰς ὥρας**, the hours
Vocative **ὥραι**, hours

NOTICE that the **α** is characteristic of the first declension, just as the **ο** is characteristic of the second declension. The genitive plural of all first declension nouns has a circumflex on the ultima (last syllable). The rules of Greek accent are available on the internet for those interested, but as mentioned before, most modern New Testament Greek textbooks do not include an explanation of accent placement.

Notice also that the dative singular form (**τῇ ὥρᾳ**) has the iota-subscript both in the definite article and in the noun itself.

43. Note the declension of **ἡ δόξα**, (the glory).

. . .

SINGULAR

Nominative **ἡ δόξα**, the glory
Genitive **τῆς δόξης**, of the glory
Dative **τῇ δόξῃ**, to/for the glory
Accusative **τὴν δόξαν**, the glory
Vocative **δόξα**, glory

PLURAL

Nominative **αἱ δόξαι**, the glories
Genitive **τῶν δοξῶν**, of the glories
Dative **ταῖς δόξαις**, to/for the glories
Accusative **τὰς δόξας**, the glories
Vocative **δόξαι**, glories

EXCEPT AFTER THE Greek letters **ε**, **ι**, and **ρ**, the **α** in the nominative singular of first declension nouns changes to **η** in the genitive and dative singular.

44. Observe the declension of **ἡ γραφή**, meaning the writing, (the) Scripture.

SINGULAR

Nominative **ἡ γραφή**, the writing
Genitive **τῆς γραφῆς**, of the writing
Dative **τῇ γραφῃ**, to/for the writing
Accusative **τὴν γραφήν**, the writing
Vocative **γραφή**, writing

. . .

PLURAL

Nominative **αἱ γραφαί** the writings
Genitive **τῶν γραφῶν**, of the writings
Dative **ταῖς γραφαῖς**, to/for the writings
Accusative **τὰς γραφάς**, the writings
Vocative **γραφαί**, writings

WHEN ANY FIRST declension noun ends in **η**, in the nominative singular, the **η** is retained throughout the singular. It will be observed that the plural forms of all first declension nouns have the same endings.

Did you notice that the endings of the nouns in the examples listed above are practically identical to the forms of their respective definite articles?

45. Translation Drill: First Declension Nouns

Translate the following Greek sentences into English. The answers are given below.

1) **ὁ μαθητὴς βλέπει τὴν δόξαν τῆς ὥρας.**
2) **ὁ μαθητὴς γινώσκει τὴν γραφήν.**
3) **ὁ προφήτης βλέπει τὰ δῶρα.**
4) **ἡ ἐκκλησία γινώσκει τὴν ἀλήθειαν.**
5) **διδάσκεις τὴν ἀγάπην τῶν διδαχῶν καὶ βλέπουσι τὴν ἐντολήν.**

1) The disciple sees the glory of the hour.
2) The disciple knows the Scripture.
3) The prophet sees the gifts.
4) The church knows the truth.

5) You teach the love of the teachings and they see the commandment.

Now translate the five English statements into Greek.

46. **Recognition Tips**

The analysis of each of the five Greek sentences in section 45, above, is given below:

1) **ὁ μαθητὴς** (masculine nominative subject), **βλέπει** (third person singular verb), **τὴν δόξαν** (feminine singular accusative [direct object]), **τῆς ὥρας** (feminine singular genitive).

2) **ὁ μαθητὴς** (masculine nominative subject), **γινώσκει** (third person singular verb), τὴν γραφήν (feminine singular accusative [direct object]).

3) **ὁ προφήτης** (masculine nominative subject), **βλέπει** (third person singular verb), **τὰ δῶρα** (neuter plural accusative [direct object]).

4) **ἡ ἐκκλησία** (feminine nominative subject), **γινώσκει** (third person singular verb), **τὴν ἀλήθειαν** (feminine singular accusative [direct object]).

5) **διδάσκεις** (second person singular nominative subject), **τὴν ἀγάπην** (feminine singular accusative [direct object]), **τῶν διδαχῶν** (feminine plural genitive), **καὶ** (conjunction = "and"), **βλέπουσι** (third person plural nominative subject), **τὴν ἐντολήν** (feminine singular accusative [direct object]).

47. **Recognition Drill**

Match the Greek with the translation. Answers are provided below. If you have difficulty with this exercise, review the grammar concepts covered thus far.

a) We have the teaching of the brothers./b) He sees the sins of the apostles./c) A man speaks to the prophets./d) The disciple knows the Scriptures./e) You see the glory of the apostles./f) You know the love and the truth of God./g) The prophet sees the gifts./h) The disciple sees the glory of the hour./i) The church knows the truth./j) You teach the love of the teachings.

_____ 1. **ἄνθρωπος λέγει τοῖς προφήταις.**
_____ 2. **ὁ μαθητὴς γινώσκει τὴν γραφήν.**
_____ 3. **ὁ μαθητὴς βλέπει τὴν δόξαν τῆς ὥρας.**
_____ 4. **βλέπει τὰς ἁμαρτίας τῶν ἀποστόλων.**
_____ 5. **ἡ ἐκκλησία γινώσκει τὴν ἀλήθειαν.**
_____ 6. **διδάσκεις τὴν ἀγάπην τῶν διδαχῶν.**
_____ 7. **βλέπεις τὴν δόξαν τῶν ἀποστόλων.**
_____ 8. **ἔχομεν τὴν διδαχὴν τῶν ἀδέλφων.**
_____ 9. **γινώσκεις τὴν ἀγάπην καὶ τὴν ἀλήθειαν θεοῦ.**
_____ 10. **ὁ προφήτης βλέπει τὰ δῶρα.**

Answers: 1c, 2d, 3h, 4b, 5i, 6j, 7e, 8a, 9f, 10g

ELEVEN

Κύριε ὁ κύριος ἡμῶν ὡς θαυμαστὸν τὸ ὄνομά σου ἐν πάσῃ τῇ γῇ.
Lord, our Lord, how majestic is your name in all the earth.
Psalm 8:9

48. **The Definite Article**

NOTE below the declension of the definite article ("the"). On the left side of the chart, the cases, "Nominative," etc., are abbreviated with a single letter. Remember that the Nominative case is the subject case, the Genitive, the possessive case, the Dative, the indirect object case, the Accusative, the direct object case and the Vocative, used for direct address. Most of the endings of the definite articles below have been encountered in the endings of the nouns of the first and second declensions presented thus far. The abbreviations M, F, and N stand for Masculine, Feminine and Neuter, respectively.

SINGULAR

N. **ὁ** (the [M]), **ἡ** (the [F]), **τό** (the [N])
G. **τοῦ** (of the [M]), **τῆς** (of the [F]),
τοῦ (of the [N])
D. **τῷ** (to/for the [M]), **τῇ** (to/for the [F]),
τῷ (to/for the [N])
A. **τόν** ([the M]), **τήν** (the [F]), **τό** (the [N])

PLURAL

N. **ὁι** (the [M]), **αἱ** (the [F]), **τά** (the [N])
G. **τῶν** (of the [M]), **τῶν** (of the [F]),
τῶν (of the [N])
D. **τοῖς** (to/for the [M]), **ταῖς** (to/for the [F]),
τοῖς (to/for the [N])
A. **τούς** (the [M]), **τάς** (the [F]), **τά** (the [N])

NOW COMPARE the forms of the definite articles listed above with the endings of the masculine, feminine and neuter nouns listed below and **recognize** the consistency in endings. The definite articles have been repeated in order to emphasize the similarities.

SINGULAR

N. **ὁ ἄνθρωπος**, the man
G. **τοῦ ἀνθρώπου**, of the man
D. **τῷ ἀνθρώπῳ**, to/for the man

A. **τὸν ἄνθρωπον**, the man
V. **ἄνθρωπε**, man

PLURAL

N. **ὁι ἄνθρωποι,** the men
G. **τῶν ἀνθρώπων**, of the men
D. **τοῖς ἀνθρώποις**, to/for the men
A. **τοὺς ἀνθρώπους**, the men
V. **ἄνθρωποι**, men

SINGULAR

N. **ἡ ὥρα**, the hour
G. **τῆς ὥρας**, of the hour
D. **τῇ ὥρᾳ**, to/for the hour
A. **τὴν ὥραν**, the hour
V. **ὥρα**, hour

PLURAL

N. **αἱ ὡραι**, the hours
G. **τῶν ὡρῶν**, of the hours
D. **ταῖς ὥραις**, to/for the hours
A. **τὰς ὥρας**, the hours
V. **ὥραι**, hours

NEUTER NOUN ENDINGS are not as consistent as masculine and feminine noun endings, but the student will note that are quite similar to the other two genders. It will be seen that in the Nominative, Accusative and Vocative, the singular and plural the forms are the same.

. . .

SINGULAR

N. **τὸ δῶρον**, the gift
G. **τοῦ δώρου**, of the gift
D. **τῷ δώρῳ**, to/for the gift
A. **τὸ δῶρον**, the gift
V. **δῶρον**, gift

PLURAL

N. **τὰ δῶρα**, the gifts
G. **τῶν δώρων**, of the gifts
D. **τοῖς δώροις**, to/for the gifts
A. **τὰ δῶρα**, the gifts
V. **δῶρα**, gifts

TWELVE

ἐξομολογήσομαί σοι κύριε ἐν ὅλῃ καρδίᾳ μου. διηγήσομαι πάντα τὰ θαυμάσιά σου.

I will praise you, O Lord, with all my heart; I will tell of all your wonders.

Psalm 9:1

Adjectives of the First and Second Declension

49. Vocabulary

ἀγαθός, ή, όν good (adjective)

δίκαιος, α, ον righteous, just, right (adjective)

ἔρημος, ἡ wilderness, desert (a "hermit" sometimes lives in the wilderness)

ἔσχατος, η, ον last (adjective) (eschatology)

ἕτερος, α, ον other, another (adjective) (heterosexual)

καινός, ή, όν new (adjective) (Cenozoic means recent geological eras)

κακός, ή, όν evil, bad, wrong, harm (adjective) (cacophony)

καλός, ή, όν good, beautiful (adjective) (calligraphy)

καθαρός, ή, όν clean (adjective) (cathartic)

μικρός, ά, όν small, little (adjective) (microscope)

μόνος, η, ον only, alone (adjective) (monorail)

ὁδός, ἡ way, road (an odometer measures the "way")

πιστός, ή, όν faithful, reliable, trustworthy (adjective)

πονηρός, ά, όν wicked, evil (adjective) (pornography)

νεκρός, ά, όν dead (adjective) (necromancy, necrophobia)

50. **Recognition Tip**

Notice that many of the Greek words listed above have cognates in English, which are given in parentheses. Remember, a cognate is a word that is very similar in spelling, pronunciation and meaning in two or more different languages.

51. Below is the declension of the Greek adjective, **ἀγαθός, ή, όν**, "good."

DO NOT BE INTIMIDATED by the adjective endings.

As mentioned before, the Greek language is endings-driven and you have already seen almost all of the endings that appear below on **ἀγαθός** on the endings of the definite articles and on the endings on the various masculine, feminine and neuter nouns in the vocabulary lists. For comparison, the declension of the definite articles is repeated below under the declension of

ἀγαθός. Thus, in most cases the definite article, the adjective and the noun will have the same endings in a Greek sentence or phrase.

SINGULAR

N. **ἀγαθός** [M] **ἀγαθή** [F] **ἀγαθόν** [N]
G. **ἀγαθοῦ** [M] **ἀγαθῆς** [F] **ἀγαθοῦ** [N]
D. **ἀγαθῷ** [M] **ἀγαθῇ** [F] **ἀγαθῷ** [N]
A. **ἀγαθόν** [M] **ἀγαθήν** [F] **ἀγαθόν** [N]

PLURAL

N. **ἀγαθοί** [M] **ἀγαθαί** [F] **ἀγαθά** [N]
G. **ἀγαθῶν** [M] **ἀγαθῶν** [F] **ἀγαθῶν** [N]
D. **ἀγαθοῖς** [M] **ἀγαθαῖς** [F] **ἀγαθοῖς** [N]
A. **ἀγαθούς** [M] **ἀγαθάς** [F] **ἀγαθά** [N]

SINGULAR

N. **ὁ** (the [M]), **ἡ** (the [F]), **τό** (the [N])
G. **τοῦ** (of the [M]), **τῆς** (of the [F]),
τοῦ (of the [N])
D. **τῷ** (to/for the [M]), **τῇ** (to/for the [F]),
τῷ (to/for the [N])
A. **τόν** ([M]), **τήν** (the [F]), **τό** (the [N])

PLURAL

N. **ὁι** (the [M]), **αἱ** (the [F]), **τά** (the [N])
G. **τῶν** (of the [M]), **τῶν** (of the [F]),
τῶν (of the [N])
D. **τοῖς** (to/for the [M]), **ταῖς** (to/for the [F]),

τοῖς (to/for the [N])
A. **τούς** (the [M]), **τάς** (the [F]), **τά** (the [N])

52. **Recognition Tip**

THE KEY to recognizing and eventually knowing the endings of both nouns and adjectives is to know the forms of the definite articles listed in the chart above. Do you see the amazing similarities when comparing the adjective endings and the definite article endings?

53. Adjectives and definite articles, which are also considered adjectives, always agree with the nouns they modify in gender and number and case. Study this example:

ὁ μαθητὴς βλέπει τὴν ἀγαθὴν δόξαν τῆς ἐσχατης ὥρας.

"The disciple sees the good glory of the last hour."

DO you recognize how the three words of the direct object singular (**τὴν ἀγαθὴν δόξαν**) are consistent in their endings? Notice also that the endings of the three words in the genitive case (**τῆς ἐσχατης ὥρας**) are also basically consistent with one another. Here is another example:

ὁ προφήτης βλέπει τὰ κακὰ δῶρα.

"The prophet sees the evil gifts."

NOTICE how the three words of the direct object plural (**τὰ κακὰ δῶρα**) are consistent in their endings and that in most cases the noun and adjective endings are the same in form as the definite article.

54. **Attributive and Predicate Use of Adjectives**

ADJECTIVES ARE EMPLOYED in three distinct ways: attributively, predicatively and substantively. Do not let these grammatical terms intimidate you.

WITH THE ATTRIBUTIVE use of the adjective, the adjective *attributes a quality to the noun* that it modifies. For example, in the phrase, *the good word,* the word *good* is an attributive adjective because it tells us something about the *word* we are talking about. We are not talking about all words or any word, but only the *good* word. In Greek there are two possible positions for the attributive adjective in relation to the noun. "The good word" can be either of these two forms:

ὁ ἀγαθὸς λογός or **ὁ λογὸς ὁ ἀγαθός**

NOTICE that the first possibility follows English exactly: "the good word." With the second possibility, one is literally saying: "the word, the good" or "the word – namely the good one."

55. **Recognition Tip**

NOTE that with the attributive construction, the adjective

comes immediately after the definite article. This fact is very important.

IN THE PREDICATE use of the adjective, we see the adjective *making an assertion* (i.e., saying something) about the noun. For example, "The word is good." In Greek two possibilities for this construction exist:

ὁ λογὸς ἀγαθός or **ἀγαθὸς ὁ λογός**

IN THIS CONSTRUCTION, the verb form "is" is not there, but is understood in each instance. For example, with **ὁ λογὸς ἀγαθός** we have, "the word (is) good," and with **ἀγαθὸς ὁ λογός** we have, "good (is) the word."

56. **Recognition Tip**

Note that with the predicate construction, the adjective is **not** immediately preceded by the definite article. This fact is very important.

THE FOLLOWING summary of these two concepts will be helpful.

Attributive position ("The good word.")

ὁ ἀγαθὸς λογός or **ὁ λογὸς ὁ ἀγαθός**

Predicate position ("The word is good.")

ὁ λογὸς ἀγαθός or **ἀγαθὸς ὁ λογός**

AN ADJECTIVE CAN BE USED as a noun. This is the substantive use. (The word "substantive" actually means "a noun.") Therefore **ὁ ἀγαθός** can mean "the good man," and **ἡ ἀγαθή** may mean "the good woman," etc.

57. **Recognition Drill**

Match the Greek with the translation. Answers are provided below.

a) She sees the righteous man./b) You see the faithful apostle./c) We have the gift of the evil brethern./d) We know the evil man./e) The good man speaks the truth./f) The prophet sees the small gifts./g) The disciple knows the other church./h) The prophet knows the only way./i) You teach the only truth./j) The disciple sees the last hours.

_____ **1. ὁ ἀγαθὸς ἄνθρωπος λέγει τὴν ἀλήθειαν.**

_____ **2. ὁ μαθητὴς γινώσκει τὴν ἕτεραν ἐκκλησίαν.**

_____ **3. ὁ μαθητὴς βλέπει τὰς ἐσχατὰς ὥρας.**

_____ **4. βλέπει τὸν δίκαιον ἄνθρωπον.**

_____ **5. ὁ προφήτης γινώσκει τὴν μόνην ὁδόν.**

_____ **6. διδάσκεις τὴν μόνην ἀλήθειαν.**

_____ **7. βλέπεις τὸν πιστὸν ἀπόστολον.**

_____ **8. ἔχομεν τὸ δῶρον τῶν πονηρῶν ἀδέλφων.**

_____ **9. γινώσκομεν τὸν κακὸν ἄνθρωπον.**

_____ **10. ὁ προφήτης βλέπει τὰ μικρὰ δῶρα.**

Answers: 1e, 2g, 3j, 4a, 5h, 6i, 7b, 8c, 9d, 10f

. . .

58. **Recognition Tips**

Did you recognize the consistencies of the endings on the definite articles, adjectives and nouns in the above sentences?

For example, in sentence 2, which features a feminine singular direct object noun, we have: **ὁ μαθητὴς γινώσκει τὴν ἕτεραν ἐκκλησίαν.** Each final **ν** has been underlined here.

In sentence 3, which features a feminine plural direct object noun, we have: **ὁ μαθητὴς βλέπει τὰς ἐσχατὰς ὥρας.** The consistencies have been underlined.

Likewise in sentence 8, which features a masculine genitive plural noun, we see: **ἔχομεν τὸ δῶρον τῶν πονηρῶν ἀδέλφων.** Again, the consistencies have been underlined.

IN OTHER WORDS, the endings of the definite articles, adjectives and nouns agree in gender, number and grammatical case, which is indicated by the endings on these words.

THIRTEEN

καὶ ἐλπισάτωσαν σὲ οἱ γινώσκοντες τὸ ὄνομά σου.
And let those who know your name, hope in you.
Psalm 9:10a

59. Additional Translation Exercises

Use a 3 X 5 card to cover the English translations, and then translate each Greek sentence into English.

1) **βλέπεις ἄνθρωπον καὶ λόγον.**
1) You see a man and a word.
2) **λόγον λέγει ἀπόστολος.**
2) An apostle says a word.
3) **βλέπει τὰ δῶρα.**
3) He (she) sees the gifts.
4) **βλέπω τὸν οἶκον.**
4) I see the house.
5) **οἱ ἄνθρωποι βλέπουσι τὸν ἀγαθὸν οἶκον.**
5) The men see the good house.

. . .

NEXT, cover the Greek sentences and translate each into English.

60. **Recognition Tips**

In sentence number 1, above, did you notice that the two singular masculine nouns (**ἄνθρωπος** and **λόγος**) became **ἄνθρωπον** and **λόγον** as direct objects? Remember, the sign of the direct object in masculine and feminine singular nouns is the Greek letter **ν** at the end of the word. The other sentences, above, feature similar concepts.

61. **Translation Exercises with the Greek Verb εἰμί**

Complete the translation practice in section 62, which features the verb **εἰμί**.

To assist you, the present active indicative of the verb, **εἰμί**, ("I am") is given below. Take a few moments to make sure you recognize the forms.

SINGULAR

1P **εἰμί** I am
2P **εἶ** you are
3P **ἐστί(ν)** he/she/it is

PLURAL

1P **ἐσμέν** we are
2P **ἐστέ** you (ye) are
3P **εἰσί(ν)** they are

. . .

62. **Translation Practice**

Translate the following English sentences into Greek. Besides giving the student practice with **εἰμί**, he or she will also receive practice with adjectives. After doing the English to Greek translations, do the reverse. Answers are provided.

1) Ye are good men.
2) He is an evil apostle.
3) We are righteous brothers.
4) You are faithful and pure.
5) They are evil and wicked.
1) **ἐστε ἀγαθοὶ ἄνθρωποι.**
2) **ἐστι κακὸς ἀπόστολος.**
3) **ἐσμεν δίκαιοι ἀδελφόι.**
4) **εἶ πιστὸς καὶ καθαρός.**
5) **εἰσι κακὸι καὶ πονηρόι.**

63. **Recognition Tip**

Once again, the endings of the definite articles, adjectives and nouns agree in gender, number and grammatical case, which is indicated by the endings on these words.

FOURTEEN

Οὗτος ἦν ἐν ἀρχῇ πρὸς τὸν Θεόν.
This One was in the beginning with God.
John 1:2

Masculine Nouns of the First Declension; Prepositions

64. This lesson introduces prepositions as well as a repetition of two masculine nouns of the first declension: **μαθητής** (disciple) and **προφήτης** (prophet). Most nouns of the first declension are feminine. The other nouns listed in the vocabulary below are nouns of the second declension. Some other vocabulary terms are repeated here from other lessons for review. Notice that many of the words have English cognates.

. . .

65. Vocabulary, prepositions, verbs and nouns.

ἀνά up, upon, again, on (adv.); thereupon (prep) (Anabaptists are baptized again.)

ἀντί (+gen) for, in place of, instead of (prep) (antibody)

ἀπό (+gen) from, by, since (prep) (An apostle is "one who is sent away.")

βάλλω I throw, cast, put (ballistic)

διά (+gen) through (prep); (+acc) because of (prep) (diameter)

εἰς (+acc) into, to, for (prep) (Eisegesis is reading one's own interpretation into a passage.)

ἐκ (+gen) of, out of, from (prep) (exit)

ἐν (+dat) in, with, by, to (prep) (Entangle means "to involve thoroughly.")

ἐπί (+gen/dat/acc) over, on, at the time of, at, to (prep) (epidermis)

θεός, ὁ God

κατά (+gen) against, down (prep); (+acc) according to (prep) (catacomb)

κοσμός, ὁ world (cosmos)

μαθητής, ὁ disciple

μένω, I remain

μετά (+gen) with (prep); (+acc) after (prep)(metamorphosis)

οὐρανός, ὁ heaven (Uranus)

παρά (+gen) from, by (prep); (+dat) with (prep); (+acc) beside (prep) (parallel)

πέμπω I send

περί (+gen) about, concerning (prep); (+acc) around (prep) (perimeter)

πρό (+gen) before, above (prep) (prototype)

πρός (+gen) for (prep); (+dat) at (prep); (+acc) to, toward (prep); moreover (adv) (Prostate means "standing before.")

προφήτης, **ὁ** prophet

σύν (+dat) with, beside (prep) (Synagogue means "together assembly.")

τέκνον, **τό,** child (technology)

τόπος, **ὁ** place (topography)

ὑπέρ (+gen) for (prep); (+acc) beyond (prep); more (adv) (hyper)

ὑπό (+gen) by, from (prep); (+acc) under (prep) (hypodermic)

66. First declension nouns ending in **-ης** are masculine. Note below the declension of **ὁ προφήτης,** the prophet.

SINGULAR

N. **ὁ προφήτης,** the prophet
G. **τοῦ προφήτου**, of the prophet
D. **τῷ προφήτῃ**, to/for the prophet
A. **τὸν προφήτην**, the prophet
V. **προφῆτα**, prophet

PLURAL

N. **ὁι προφῆται,** the prophets
G. **τῶν προφητῶν**, of the prophets
D. **τοῖς προφήταις**, to/for the prophets
A. **τοὺς προφήτας**, the prophets
V. **προφήτας**, prophets

67. **Recognition Tip**

. . .

NOTE that despite the fact that **ὁ προφήτης** is masculine, it adheres to the first declension noun pattern, being exactly like a first declension noun except in the nominative, genitive and vocative.

The noun **μαθητής** is declined like **προφήτης**, except the accents are different. Do not worry about the placements of accents marks. They were added to the text centuries after the original manuscripts were created.

68. **Prepositions**

A PREPOSITION EXPRESSES RELATIONSHIP. In the sentence, "The man is in the temple," the preposition "in" expresses and defines a certain relationship between the man and the temple. In the sentence, "The man goes into the temple," a different preposition, "into" expresses a different relationship.

In English the relationship between prepositions and nouns is relatively simple. Nouns and pronouns coming after prepositions are always in the "objective" (accusative) case. In Greek, however, different prepositions take different cases. In the vocabulary list above, +acc means that the preposition takes the accusative case, whereas +dat indicates that the preposition takes the dative case and +gen means that the preposition takes the genitive case.

PREPOSITIONS WHICH END in a vowel (except **περί** and **πρό**) drop the vowel when the next word begins with a vowel. Note that an apostrophe then replaces the last vowel in the preposition except in compound verb forms such as **ἀπό ἄγω**, which becomes **ἀπάγω**. (I depart.)

69. **Recognition Tip**

WE WILL NOW DISCOVER that three of the grammatical cases we have studied have other functions. The accusative case, indeed, includes direct objects, but is also required after certain prepositions. The dative does include indirect objects, but is required after certain prepositions. Likewise, the genitive case includes concepts of possession, but is also required after certain prepositions.

Examples:

The preposition **ἐν**, meaning "in," always takes the dative case. Thus, the Greek equivalent of the English the sentence, "in the house" is **ἐν τῷ οἴκῳ**.

The preposition **εἰς**, which means "into," always takes the accusative case. Therefore, "into the house" must be expressed in Greek as **εἰς τὸν οἶκον.**

The preposition **ἀπό** takes the genitive case. So, "from the house" in Greek is **ἀπὸ τοῦ οἴκου.**

70. **Recognition Tip**

Note that the three examples above define a general principle:

The accusative case is the case of "motion toward" a place or thing.

The dative case is the case of "rest" in a place.

The genitive case is the case of "separation" from a place, person, etc.

Therefore, prepositions expressing motion toward a place, a person, etc., normally take the accusative case, prepositions denoting rest in a place usually take the dative case and prepositions expressing separation from a place, person, etc., take the genitive case.

Please note that these are general rules. Many prepositions do not adhere to these general principles. Therefore, for example, some prepositions which do not express any idea of separation take the genitive case, nonetheless.

Note that the prepositions **εἰς**, **ἐν**, and **ἐκ** (out of) are proclitics. A proclitic is a word that goes so closely with the following word as to have no accent of its own.

Three prepositions we have mentioned, **εἰς**, **ἐν**, and **ἀπό**, take only one case. Another pronoun **πρός,** "to," almost always takes the accusative case.

However, there are a number of prepositions which take the dative in one instance, the accusative in another, etc., depending on the meaning the author wants to express. In addition, prepositions which take several cases may have quite a different meaning in the accusative compared to the dative. The preposition **διά** is a case in point. In the genitive **διά** means "through" and in the accusative the same preposition means "on account of." Let's consider another preposition: **μετά**. In the accusative it means "after" and with the genitive it means "with."

DO NOT SIMPLY LEARN what the prepositions mean. Recognize with what case they are associated in a specific situation. To quote Gresham Machen, in "New Testament Greek for Beginners," "Thus it is quite insufficient to say that **ἐν** means 'in.' What should rather be said is that '**ἐν** -with-the-dative means in.' The phrase '**ἐν** -with-the-dative' should form in the student's mind one absolutely indivisible idea; **ἐν** should never be thought of apart from its case. In the same way, but still more obviously, it is insufficient to say that **μετά** means 'with' or 'after.' What should rather be said is that '**μετά** -with-the-genitive means with, and that **μετά** -with-the-accusative means after.' This same method of study should be applied to all prepositions."

Please note that the meanings given in Greek grammars for prepositions are the usual meanings of these words. In some instances **ἐν** cannot be translated by the English word "in." However, for present purposes, the student should adhere to the meanings given above.

71. **Recognition Tips**

Preposition Summary

GENERAL CASE CONCEPTS IN GREEK:

Accusative Case: Motion towards a person or object

Dative Case: Location in or at a place; rest in or at a place

Genitive Case: Motion through or away from a person or place; possession

GREEK PREPOSITIONS which take only the accusative:

εἰς into, to, for

πρός to, toward; (adverb) moreover

PREPOSITIONS TAKING ONLY THE DATIVE:

ἐν in, with, by, to

σύν with, beside

PREPOSITIONS TAKING ONLY THE GENITIVE:

ἀντί for, in place of, instead of (antibody)
ἀπό from, by, since
ἐκ of, out of, from
πρό before, above (prototype)

PREPOSITIONS TAKING the accusative and genitive:

διά (+gen) through; (+acc) because of (diameter)
κατά (+gen) against, down; (+acc) according to (catacomb)
μετά (+gen) with; (+acc) after (metamorphosis)
περί (+gen) about, concerning; (+acc) around (perimeter)
ὑπέρ (+gen) for; (+acc) beyond; (adv) more (hyper)
ὑπό (+gen) by, from; (+acc) under (hypodermic)

PREPOSITIONS TAKING ALL THREE CASES:

ἐπί (+gen/dat/acc) over, on, at the time of, at, to (epidermis)

παρά (+gen) from the side of, by; (+dat) with; (+acc) beside (parallel)

72. **Greek Verses Containing Prepositions**

Greek prepositions, as well as the English equivalents, are bolded and underlined.

MARK 1:23

Καὶ ἦν **ἐν** τῇ συναγωγῇ αὐτῶν ἄνθρωπος
ἐν πνεύματι ἀκαθάρτῳ.
And immediately there was **in** their
synagogue a man **with** an unclean spirit.

(Note that the two nouns in the dative singular (συναγωγῇ and πνεύματι) end in **ι**. The sign of the dative singular in all declensions is this final **ι**. With συναγωγῇ the final **ι** is under η as ῃ. This concept works most of the time. An exemption would be οἱ ἄνθρωποι (the men), but here both words are in the nominative plural, not the dative singular.

MARK 1:29

Καὶ εὐθέως **ἐκ** τῆς συναγωγῆς ἐξελθόντες
ἦλθον **εἰς** τὴν οἰκίαν Σίμωνος καὶ
᾿Ανδρέου **μετὰ** ᾿Ιακώβου καὶ ᾿Ιωάννου.
And as soon as they came **out of** the
synagogue, they entered **into** the house of
Simon and Andrew **with** James and John.

JOHN 8:23

καὶ εἶπεν αὐτοῖς· ὑμεῖς **ἐκ** τῶν κάτω ἐστέ,
ἐγὼ **ἐκ** τῶν ἄνω εἰμί· ὑμεῖς **ἐκ** τοῦ κόσμου
τούτου ἐστέ, ἐγὼ οὐκ εἰμὶ **ἐκ** τοῦ κόσμου
τούτου.
And he said to them, "You are **from** beneath
and I am **from** above. You are **from**
this world, I am not **from** this world. (Do you see that **ἐκ** takes the genitive case? In the first two instances it is the genitive plural and in the last two occurrences it is the genitive singular.)

. . .

JOHN 9:17

λέγουσι τῷ τυφλῷ πάλιν· σὺ τί λέγεις **περὶ** αὐτοῦ, ὅτι ἤνοιξέ σου τοὺς ὀφθαλμούς; ὁ δὲ εἶπεν ὅτι προφήτης ἐστίν.

Therefore, they say again to the blind man: "What do you say **concerning** him, because he opened your eyes?" And he said, "He is a prophet."

1 JOHN 4:7

Αγαπητοί, ἀγαπῶμεν ἀλλήλους, ὅτι ἡ ἀγάπη **ἐκ** τοῦ Θεοῦ ἐστι, καὶ πᾶς ὁ ἀγαπῶν **ἐκ** τοῦ Θεοῦ γεγέννηται καὶ γινώσκει τὸν Θεόν.

Beloved, let us love one another because love comes **from** God, and everyone who loves is born **of** God and knows God.

1 JOHN 4:18

φόβος οὐκ ἔστιν **ἐν** τῇ ἀγάπῃ, ἀλλ᾽ ἡ τελεία ἀγάπη ἔξω βάλλει τὸν φόβον, ὅτι ὁ φόβος κόλασιν ἔχει.

There is no fear **in** love, but perfect love casts out fear, because fear involves torment.

73. Translation Exercises with Greek Prepositions

Translate the following English sentences into Greek. After doing that, cover the Greek sentences and translate them into English. Answers are provided below.

1) The man casts the evil apostle out of the house.

2) The messenger of the apostle comes to him.

3) The brother is with his servant.

4) He comes into the small house with his servant.

1) ὁ ἄνθρωπος βάλλει τὸν πονηρὸν ἀπόστολον ἐκ τοῦ οἴκου.

2) ὁ ἄγγελος τοῦ ἀποστόλου ἔρχεται πρὸς αὐτὸν.

3) ὁ ἀδελφὸς ἐστιν σὺν τῷ δούλῳ αὐτοῦ.

4) ἔρχεται εἰς τὸν μικρὸν οἶκον σὺν τῷ δούλῳ αὐτοῦ.

74. **Recognition Drill.**

Match the Greek with the translation. Answers are provided below. (Notice the agreement of definite articles, adjectives and nouns.)

a) There is no fear in love./b) The man comes out of the house./c) I come around the house./d) Love is from God./e) The apostle comes into the house./f) He is in the synagogue./g) I am in the good temple./h) He speaks the words to the men./i) The gifts are under the house./j) The gifts are with the brother.

_____ **1. ὁ ἄνθρωπος ἔρχεται ἐκ τοῦ οἴκου.**

_____ **2. ὁ ἀπόστολος ἔρχεται εἰς τὸν οἶκον.**

_____ **3. εἰμὶ ἐν τῷ ἀγαθῷ ἱερῷ.**

_____ **4. λέγει τοὺς λόγους τοῖς ἀνθρώποις.**

_____ **5. ἔρχομαι περὶ τὸν οἶκον.**

_____ **6.τὰ δῶρα εἰσὶν ὑπὸ τὸν οἶκον.**

_____ **7. τὰ δῶρα εἰσὶν σὺν τῷ ἀδέλφῳ.**

_____ **8. ἔστιν ἐν τῇ συναγωγῇ.**

_____ **9. ἡ ἀγάπη ἐκ τοῦ Θεοῦ ἐστι.**

_____ **10. φόβος οὐκ ἔστιν ἐν τῇ ἀγάπῃ.**

Answers: 1b, 2e, 3g, 4h, 5c, 6i, 7j, 8f, 9d, 10a

FIFTEEN

ἐν τῷ κόσμῳ ἦν, καὶ ὁ κόσμος δι᾽ αὐτοῦ ἐγένετο, καὶ ὁ κόσμος αὐτὸν οὐκ ἔγνω.
He was in the world, and the world was made through Him, and the world did not know Him.
John 1:10

Personal Pronouns

75. The declension of the personal pronoun of the first person is given below. Notice that the genitive singular form ends in **ου** like most masculine nouns in the genitive singular. The dative singular ends in **ι** like all masculine nouns in the dative singular.

In addition, the student will note that the genitive plural ends in **ων** like all nouns as well as all definite articles in the genitive plural. Knowing these facts will aid in recognition.

. . .

SINGULAR

N. **ἐγώ** (I)
G. **ἐμοῦ** or **μου** (of me)
D. **ἐμοί** or **μοι** (to/for me)
A. **ἐμέ** or **με** (me)

PLURAL

N. **ἡμεῖς** (we)
G. **ἡμῶν** (of us)
D. **ἡμῖν** (to/for us)
A. **ἡμᾶς** (us)

The forms **ἐμοῦ**, **ἐμοί**, and **ἐμέ** are emphatic forms; they are used when emphasis is desired. The unemphatic forms of these three words (**μου**, **μοι**, **με**) are called enclitic and have no accents.

76. The declension of the personal pronoun of the second person is as follows. Notice again that the genitive singular ends in **ου** like most masculine nouns in the genitive singular. The dative singular ends in **ι** like masculine nouns in the dative singular. In addition, the genitive plural ends in **ων** like all nouns in the genitive plural.

SINGULAR

N. **σύ** (thou [you])
G. **σοῦ** or **σου** (of thee [you])
D. **σοί** or **σοι** (to/for thee [you])
A. **σέ** or **σε** (thee [you])

. . .

PLURAL

N. **ὑμεῖς** (ye [you])

G. **ὑμῶν** (of you)

D. **ὑμῖν** (to/for you)

A. **ὑμᾶς** (you)

The forms **σοῦ**, **σοί**, and **σέ** are emphatic forms; they are used when emphasis is desired. The unemphatic forms of these three words (**σου**, **σοι**, and **σε)** are enclitic and have no accents.

77. The declension of the personal pronoun of the third person is given below. The endings are almost identical to those of the definite article and of the adjective, **ἀγαθός.**

SINGULAR

N. **αὐτός** (he) **αὐτή** (she) **αὐτό** (it)

G. **αὐτοῦ** (of him) **αὐτῆς** (of her) **αὐτοῦ** (of it)

D. **αὐτῷ** (to/for him) **αὐτῇ** (to/for her) **αὐτῷ** (to/for it)

A. **αὐτόν** (him) **αὐτήν** (her) **αὐτό** (it)

PLURAL

N. **αὐτόι** (they) **αὐταί** (they) **αὐτά** (they)

G. **αὐτῶν** (of them) **αὐτῶν** (of them) **αὐτῶν** (of them)

D. **αὐτοῖς** (to/for them) **αὐταῖς** (to/for

them) **αὐτοῖς** (to/for them)
A. **αὐτούς** (them) **αὐτάς** (them) **αὐτά** (them)

78. A pronoun is a word that replaces or stands for a noun. For instance, the sentence, "I know the man and teach him," has the same meaning as "I know the man and teach the man." In the first sentence, the pronoun him appears in the place of the second occurrence of the word "man."

The noun for which a pronoun stands is called its antecedent. Therefore, in the sentence, "I know the man and teach him," the antecedent of "him" is "the man."

A PRONOUN AGREES with its antecedent in gender and number and case.

Example: **γινώσκουσιν τὸν λογὸν καὶ γράφουσιν αὐτόν**, means, "They know the word and they write it." In this case **λογόν** is the antecedent of **αὐτόν** and since **λογόν** is of masculine gender and singular number and accusative case, **αὐτόν** is also masculine singular accusative. Notice that **αὐτόν** in this case means "it" referring to the "word" and not to a male animal or person.

PERSONAL PRONOUNS ARE NOT USED in the nominative case unless they are used for emphasis. This is what we mean by the emphatic use. **βάλλω** sufficiently translates as "I cast." **ἐγὼ βάλλω** would mean "**I** cast." (Emphasis on "I")

. . .

79. **Recognition Tip**

IMPORTANT FACT. Please note how Greek forms ownership using the genitive.

"Your house in Greek" is **ὁ οἶκός σου**, which literally means, "The house of you." Similarly, **ὁ οἶκος αὐτοῦ** literally means "The house of him," but we would translate this phrase as "His house."

80. **Sample Greek Verses Containing Personal Pronouns**

The pronouns, as well as the English equivalents, are bolded and underlined.

MARK 2:11

σοὶ λέγω, ἔγειρε καὶ ἆρον τὸν κράβαττόν **σου** καὶ ὕπαγε εἰς τὸν οἶκόν **σου**.

To you I say, "Rise up and take **your** bed and go home. (The literal meaning of this verse is, "Rise up and take the bed **of you** and go to the home **of you**).

MARK 3:34

καὶ περιβλεψάμενος κύκλῳ τοὺς περὶ αὐτὸν καθημένους λέγει· ἴδε ἡ μήτηρ **μου** καὶ οἱ ἀδελφοί **μου**.

And looking around at those sitting around him, he says, "Behold **my** mother and **my** brothers.

. . .

JOHN 2:4

λέγει αὐτῇ ὁ Ἰησοῦς· τί **ἐμοὶ** καὶ **σοί**, γύναι; οὔπω ἥκει ἡ ὥρα **μου**.

And Jesus says to her, "Why does this concern **me** or **you**, mother? **My** hour has not yet come."

JOHN 6:55

ἡ γὰρ σάρξ **μου** ἀληθῶς ἐστι βρῶσις, καὶ τὸ αἷμά **μου** ἀληθῶς ἐστι πόσις.

For **my** flesh is true food. And **my** blood is true drink.

JOHN 9:9–10

ἐκεῖνος ἔλεγεν ὅτι ἐγώ εἰμι. ἔλεγον οὖν **αὐτῷ**· πῶς ἀνεῴχθησάν **σου** οἱ ὀφθαλμοί;

That man was saying, "I am he." They were saying therefore **to him**, "How then were **your** eyes opened?"

1 JOHN 2:26

Ταῦτα ἔγραψα **ὑμῖν** περὶ τῶν πλανώντων ὑμᾶς.

I wrote these things **to you** concerning the ones who are deceiving you.

. . .

81. **Translation Exercises with Greek Personal Pronouns**

Translate the following English sentences into Greek. After doing the English to Greek translations, cover the Greek sentences and translate them into English. Answers are provided below.

1) To you I say, go (ὕπαγε) into your house.

2) She sees me and my brothers.

3) My brother sees his gift.

4. His mother says to him, "Do you see your house?"

1) σοὶ λέγω, ὕπαγε εἰς τὸν οἶκόν σου.

2) βλέπει με καὶ τοὺς ἀδελφούς μου. ·

3) ὁ ἀδελφός μου βλέπει τὸ δῶρον αὐτοῦ.

4) ἡ μήτηρ αὐτοῦ λεγει αὐτῷ, βλέπεις τὸν οἶκόν σου;

82. **Recognition Drill**

Match the Greek with the translation. Answers are provided below.

a) His gifts are with the brother./b) I am in her synagogue./c) I come around my house./d) You see your house./e) He speaks my words./f) He knows the word and he writes it./g) Your love is good./h) He is in our synagogue./i) I am in their temple./j) Your gifts are under the house.

_____ **1. γινώσκει τὸν λογὸν καὶ γράφει αὐτόν.**

_____ **2. βλέπεις τὸν οἶκόν σου.**

_____ **3. εἰμὶ ἐν τῷ ἱερῷ αὐτῶν.**

_____ **4. λέγει τοὺς λόγους μου.**

_____ **5. ἔρχομαι περὶ τὸν οἶκον μου.**

_____ **6.τὰ δῶρα σου εἰσὶν ὑπὸ τὸν οἶκον.**

_____ 7. τὰ δῶρα αὐτοῦ εἰσὶν σύν τῷ ἀδέλφῳ.
_____ 8. εἰμὶ ἐν τῇ συναγωγῇ αὐτῆς.
_____ 9. ἡ ἀγάπη ὑμῶν ἐστὶν καλή.
_____ 10. ἔστὶν ἐν τῇ συναγωγῇ ἡμῶν.

Answers: 1f, 2d, 3i, 4e. 5c, 6j, 7a, 8b, 9g, 10h

SIXTEEN

Καὶ ὁ Λόγος σὰρξ ἐγένετο καὶ ἐσκήνωσεν ἐν ἡμῖν.
And the Word became flesh and dwelt among us.
John 1:14a

83. **Recognition Review: More Greek verses featuring the verb, εἰμι (to be)**

Applicable verbs, as well as the English equivalents are bolded and underlined. For practice, some verses are repeated from previous lessons.

JOHN 15:1

Ἐγώ **<u>εἰμι</u>** ἡ ἄμπελος ἡ ἀληθινή, καὶ ὁ
πατήρ μου ὁ γεωργός **<u>ἐστι</u>**.
I **<u>am</u>** the true vine and my father **<u>is</u>** the
vinedresser.

. . .

JOHN 15:3

ἤδη ὑμεῖς καθαροί **ἐστε** διὰ τὸν λόγον ὃν λελάληκα ὑμῖν.

You **are** already clean because of the word which I have spoken to you.

JOHN 1:49

ἀπεκρίθη Ναθαναὴλ καὶ λέγει αὐτῷ, ῥαββί, σὺ **εἶ** ὁ υἱὸς τοῦ Θεοῦ, σὺ **εἶ** ὁ βασιλεὺς τοῦ Ἰσραήλ.

Nathanael answered and says to him, "Rabbi, you **are** the son of God, you **are** the king of Israel.

JOHN 14:9

λέγει αὐτῷ ἡ γυνή, Κύριε, θεωρῶ ὅτι προφήτης **εἶ** σύ.

The woman says to him, "Lord, I see that you **are** a prophet."

JOHN 6:20

ὁ δὲ λέγει αὐτοῖς· ἐγώ **εἰμι**· μὴ φοβεῖσθε.

And he says to them, "It **is** I; do not fear."

JOHN 6:48

ἐγώ **εἰμι** ὁ ἄρτος τῆς ζωῆς.

I **am** the bread of life.

. . .

JOHN 6:55

ἡ γὰρ σάρξ μου ἀληθῶς **<u>ἐστι</u>** βρῶσις, καὶ τὸ αἷμά μου ἀληθῶς **<u>ἐστι</u>** πόσις.

For my flesh **<u>is</u>** truly food, and my blood **<u>is</u>** truly drink.

1 JOHN 5:7A

ὅτι τρεῖς **<u>εἰσιν</u>** οἱ μαρτυροῦντες ἐν τῷ οὐρανῷ, ὁ Πατήρ, ὁ Λόγος καὶ τὸ Ἅγιον Πνεῦμα.

Because there **<u>are</u>** three who bear witness in heaven, the Father and the Word and the Holy Spirit.

SEVENTEEN

Ἴδε ὁ Ἀμνὸς τοῦ Θεοῦ ὁ αἴρων τὴν ἁμαρτίαν τοῦ κόσμου.
Behold the Lamb of God, Who takes away the sin of the world.
John 1:29b

Demonstrative Pronouns; Further use of αὐτός

84. Vocabulary

ἀγάπη, ἡ love
ἁμαρτία, ἡ sin
βαπτίζω I baptize
διδάσκολος, ὁ teacher (didactic)
ἐκεῖνος, η, ο that, that one (pronoun)
ἐπαγγελία, ἡ promise
εὐαγγέλιον, τό gospel (evangelize)

καλός, ά, όν, good (similar in meaning to **ἀγαθός**) (Calligraphy is "good writing.")

κρίνω I judge

νῦν now (adverb) (**νῦν** sounds like the English word "now.")

οὗτος, αὕτη, τοῦτο this, this one (pronoun)

οὕτως thus, so (adverb)

πονηρπός, ά, όν evil (adjective) (porno)

πρόσωπον, τό face

χαρά, ἡ joy

REMEMBER that when nouns such as **ἀγάπη** or **ἁμαρτία** are used without the definite article (**ἡ** in this case), the meaning would usually be:

ἀγάπη a love, love/ **ἁμαρτία** sin, a sin

85. The declension of the demonstrative pronoun **οὗτος**, "this/this one," is given below. The plural forms would, of course, mean "these," etc.

The student should find the endings on these pronouns quite familiar.

SINGULAR

N. **οὗτος** (this [M]), **αὕτη** (this [F]), **τοῦτο** (this [N])

G. **τούτου** (of this [M]), **ταὐτης** (of this [F]), **τούτου** (of this [N])

D. **τούτῳ** (to/for this [M]), **ταύτῃ** (to/for this [F]), **τούτῳ** (to/for this [N])

A. **τοῦτον** (this [M]), **ταύτην** (this [F]), **τοῦτο** (this [N])

. . .

PLURAL

N. **οὗτοι** (these [M]), **αὗται** (these [F]), **ταῦτα** (these [N])

G. **τούτων** (of these [M]), **τούτων** (of these [F]), **τούτων** (of these [N])

D. **τούτοις** (to/for these [M]), **ταύταις** (to/for these [F]), **τούτοις** (to/for these [N])

A. **τούτους** (these [M]) **ταύτας** (these [F]), **ταῦτα** (these [N])

THE DECLENSION of the demonstrative pronoun **ἐκεῖνος**, meaning "that," or "that one," is almost exactly like that of **οὗτος** except that the accent marks are slightly different.

86. **Recognition Tip**

Do you see that the endings of **οὗτος**, and **αὐτός** (covered in chapter fifteen) are almost identical to the forms of the Greek definite articles?

87. **The uses of οὗτος and ἐκεῖνος**

Study these examples:

"This word" = **οὗτος ὁ λόγος** or **ὁ λόγος οὗτος**

"That word" = **ἐκεῖνος ὁ λόγος** or **ὁ λόγος ἐκεῖνος**

"He (she, it) sees the word." = **βλέπει τὸν λόγον.**

"He (she, it) sees this word." = **βλέπει τοῦτον τὸν λόγον.** or **βλέπει τὸν λόγον τοῦτον.**

"This good word" = **οὗτος ὁ καλὸς λόγος** or **ὁ καλὸς λόγος οὗτος**

οὗτος and ἐκεῖνος are often used with nouns. When they are used in this manner, the noun with which they are used has the definite article and **οὗτος** or **ἐκεῖνος** are in the predicate position, so that **οὗτος ὁ λόγος** or **ὁ λόγος οὗτος** is like saying "this the word" or "the word this," respectively. However, both phrases have the same English meaning of "this word."

οὗτος and **ἐκεῖνος** can be used by themselves, as nouns.

Examples:

οὗτος (this man/this person), **αὕτη** (this woman), **τοῦτο** (this thing), **οὗτοι** (these men), **αὗται** (these women), **ταῦτα** (these things).

88. **Further use of αὐτός**

The primary uses of the adjectives and the pronouns studied thus far, along with two new uses of **αὐτός**, are:

"The good word" = **ὁ καλὸς λόγος** or **ὁ λόγος ὁ καλός**

"The word is good" = **καλὸς ὁ λόγος** or **ὁ λόγος καλός**

"This word" = **οὗτος ὁ λόγος** or **ὁ λόγος οὗτος**

"That word" = **ἐκεῖνος ὁ λόγος** or **ὁ λόγος ἐκεῖνος**

"The word itself" = **αὐτὸς ὁ λόγος** or **ὁ λόγος αὐτός**

"The same word" = **ὁ αὐτὸς λόγος** or **ὁ λόγος ὁ αὐτός**

"My word" = **ὁ λόγος μου**

"His word" = **ὁ λόγος αὐτοῦ**

"I see this man." = **βλέπω τοῦτον.**

"Í see this word." = **βλέπω τοῦτον τὸν λόγον.**

"You (thou) see these things." = **βλέπεις ταῦτα.**

89. **Recognition Drill**

Match the Greek with the translation. Answers are provided below.

a) This word is good./b) His word is evil./c) You see this good word./d) He sees these things./e) I see this word./f) She sees this man./g) We see this man./h) That word is good./i) My word is evil./j) That teacher is good.

Notice the agreement of definite articles, adjectives and nouns.

_____ **1. ὁ λόγος μου ἐστὶν πονηρπός.**

_____ **2. ἐκεῖνος ὁ διδάσκολος ἐστὶν ἀγαθός.**

_____ **3. βλέπει ταῦτα.**

_____ **4. βλέπομεν τοῦτον.**

_____ **5. οὗτος ὁ λόγος ἐστὶν ἀγαθός.**

_____ **6. ὁ λόγος αὐτοῦ ἐστὶν πονηρπός.**

_____ **7. βλέπεις τοῦτον τὸν ἀγαθὸν λόγον.**

_____ **8. βλέπω τοῦτον τὸν λόγον.**

_____ **9. βλέπει τοῦτον τὸν ἄνθρωπον.**

_____ **10. ἐκεῖνος ὁ λόγος ἐστὶν ἀγαθός.**

Answers: 1i, 2j, 3d, 4g, 5a, 6b, 7c, 8e, 9f, 10h

EIGHTEEN

Οὗτός ἐστιν ὁ Υἱὸς τοῦ Θεοῦ.
This is the Son of God.
John 1:34b

Present Middle and Passive Indicative; Deponent Verbs

90. Vocabulary

ἀλλά but (conjunction), stronger than **δέ**

ἀκούω I hear; (takes the genitive case and sometimes the accusative) (acoustic)

ἁμάρτωλος, ὁ sinner

ἀποκρίνομαι I answer (takes the dative case); deponent verb

ἄρχω I rule (take the genitive case); the middle form (**ἄρχομαι**) means, "I begin."

γίνομαι I become; I am (takes the nominative case); deponent verb

διέρχομαι I go through; deponent verb

εἰσέρχομαι I go in, enter; deponent verb

ἐξέρχομαι I go out; deponent verb

ἔρχομαι I come, go; deponent verb

ὅτι that, because (conjunction)

οὐ not (**οὐκ** before smooth-breathing vowels), **οὐχ** before rough-breathing vowels. For example, the **ἀ** in **ἀλλά** is smooth-breathing because the breathing mark over it slopes to the left, whereas in the case of **ἁμάρτωλος** the **ἁ** is rough-breathing because the breathing mark slopes to the right.

πορεύομαι I go; deponent verb

σώζω I save

ὑπό by (preposition) with the genitive case; under with the accusative case (hypodermic)

91. The active and passive voices are used in Greek in the same manner they are used in English. (See Chapter Five for the discussion of the concept of voice.)

IN THE MIDDLE voice the subject acts in some fashion that concerns itself, or acts upon something that belongs to itself. Much of the time it is impossible to indicate any difference in an English translation between the Greek active and the middle voice.

In other instances, the difference between the active and middle forms of a verb is radical. For example: **ἄρχω** in the active voice means "I rule" and in the middle voice (**ἄρχομαι**) it means "I begin."

. . .

ALTHOUGH THE MIDDLE voice of **λύω** (I loose) does not occur in the New Testament, the forms given below will help the student recognize the forms of the Greek verbs that do appear in the middle voice. The forms of the Present Middle Indicative and the Present Passive Indicative are given below. The forms for both of these are the same, but the meanings are quite different.

With practice the student will learn to sort out the correct meaning in a particular verse.

92. Present Middle Indicative of **λύω**

SINGULAR

1P **λύομαι** I loose (am loosing) for myself
2P **λύῃ** you loose (are loosing) for yourself
3P **λύεται** he looses (is loosing) for himself

PLURAL

1P **λυόμεθα** we loose (are loosing) for ourselves
2P **λύεσθε** you loose (are loosing) for yourselves
3P **λύονται** they loose (are loosing) for themselves

IN THESE FORMS, as in the passive forms listed below, the stem of the verb (**λυ**-) is taken and the following primary tense middle/passive endings are added: -**μαι, -σαι, -ται, -μεθα, -σθε, -νται**.

To facilitate pronunciation a variable vowel is placed between the stem of the verb and the ending. The letter **ο** is inserted before an ending that begins with **μ** or **ν**, and **ε** is placed before other letters. The second person singular (**λύῃ**) is a shortened form instead of **λύεσαι**.

PLEASE NOTE that the present active middle forms of verbs (given above), where the action of the verb is directed back at the subject, are fairly rare. The student will see these forms used most often with so-called deponent verbs, which are explained in detail below.

93. The Present Passive Indicative of **λύω** is below. Please carefully note that these forms are exactly the same as the Present Middle Indicative forms, above, but have different meanings and uses.

SINGULAR

1P **λύομαι** I am being loosed
2P **λύῃ** you are being loosed
3P **λύεται** he is being loosed

PLURAL

1P **λυόμεθα** we are being loosed
2P **λύεσθε** you are being loosed
3P **λύονται** they are being loosed

94. **Recognition Tip**

Let us consider a typical Greek verb **σώζω**, (I save) in the present active indicative and in the present passive indicative forms. The present active indicative form is **σώζω**, "I save."

The present passive indicative form of this verb is **σῴζομαι**, which means "I am being saved."

95. The use of ὑπό (by) with the genitive

ὑπό can take the accusative, with a meaning of "under," but this preposition is most often used in the genitive. With that usage, it expresses the agent by which an act is performed, and this use occurs most often in the passive voice.

Consider the following two scenarios:

σῴζει με. "He saves me." (present active indicative)

σῴζομαι ὑπ' αὐτοῦ. "I am being saved by him." The **ό** of **ὑπό** is replaced by an apostrophe when **ὑπό** comes before a word that begins with a vowel, in this case it was **αὐτοῦ**.

In the second sentence we are expressing the same thought in the passive voice as we expressed in the active voice of the first sentence. In the second sentence the object of the active verb **σῴζει** becomes the subject of the passive and the subject of the verb in the first sentence ("he saves") becomes **ὑπό** with the genitive.

96. Deponent Verbs

Deponent verbs are those which have no active forms, but only middle or passive forms, but are active in meaning. For example:

ἔρχομαι ("I come.") is passive in form like **λύομαι** ("I am being loosed."), but **ἔρχομαι** (meaning "I come") is active in meaning like **λύω** ("I loose").

97. The position of οὐ (not)

The negative particle **οὐ** precedes the word it negates, which is usually a verb. For example:

οὐ λύομαι means "I am not being loosed."

98. **Various Cases with Verbs**

Many Greek verbs take the dative or genitive case to complete their meanings when the corresponding English verb takes a direct object (accusative case). For example:

ἀκούω τῆς φωνῆς. "I hear the voice." Genitive case.

ἀποκρίνομαι τῇ φωνῇ. "I answer the voice." Dative case.

99. **Greek Verses Featuring Deponent Verbs**

Deponent verbs, as well as the English equivalents, are bolded and underlined.

MARK 3:31

Καὶ **<u>ἔρχεται</u>** ἡ μήτηρ αὐτοῦ καὶ οἱ ἀδελφοὶ
αὐτοῦ καὶ ἔξω στήκοντες ἀπέστειλαν πρὸς
αὐτὸν καλοῦντες αὐτόν.
And **<u>comes</u>** his mother and his brothers
and (while) standing outside, they sent
(word) to him, summoning him.

MARK 5:38A

Καὶ **<u>ἔρχονται</u>** εἰς τὸν οἶκον τοῦ
ἀρχισυναγώγου.
And **<u>they come</u>** into the house of the leader
of the synagogue.

. . .

MARK 3:20A

Καὶ **<u>ἔρχεται</u>** εἰς οἶκον καὶ **<u>συνέρχεται</u>** πάλιν [ὁ] ὄχλος.

And **<u>he comes</u>** into a house and (the) crowd **<u>gathers</u>** (there) again.

MARK 1:21

Καὶ **<u>εἰσπορεύονται</u>** εἰς Καπερναούμ· καὶ εὐθέως τοῖς σάββασιν εἰσελθὼν εἰς τὴν συναγωγὴν ἐδίδασκε.

And **<u>they enter</u>** into Capernaum. And immediately on the Sabbath, entering the synagogue, he was teaching.

MARK 8:22

Καὶ **<u>ἔρχονται</u>** εἰς Βηθσαϊδάν. καὶ φέρουσιν αὐτῷ τυφλὸν καὶ παρακαλοῦσιν αὐτὸν ἵνα αὐτοῦ ἅψηται.

And **<u>they enter</u>** into Bethsaida. And they bring to him a blind man and entreaty him, that he might touch him.

MARK 11:27

Καὶ **<u>ἔρχονται</u>** πάλιν εἰς Ἱεροσόλυμα. καὶ ἐν τῷ ἱερῷ περιπατοῦντος αὐτοῦ **<u>ἔρχονται</u>** πρὸς αὐτὸν οἱ ἀρχιερεῖς καὶ οἱ γραμματεῖς καὶ οἱ πρεσβύτεροι.

And **<u>they enter</u>** again into Jerusalem. And,

as he is walking around in the temple, the chief priests and the scribes and the elders **come to** him.

MARK 14:37A

καὶ **ἔρχεται** καὶ εὑρίσκει αὐτοὺς καθεύδοντας, καὶ λέγει τῷ Πέτρῳ, Σίμων, καθεύδεις;

He **comes** and finds them sleeping, and said to Peter, "Simon, are you sleeping?"

MARK 1:40A

Καὶ **ἔρχεται** πρὸς αὐτὸν λεπρὸς παρακαλῶν αὐτὸν καὶ γονυπετῶν.

And a leper **comes** to him, imploring him and kneeling.

1 JOHN 2:18A

Παιδία, ἐσχάτη ὥρα ἐστί, καὶ καθὼς ἠκούσατε ὅτι ὁ ἀντίχριστος **ἔρχεται**, καὶ νῦν ἀντίχριστοι πολλοὶ γεγόνασιν.

Little children, it is the last hour, and just as you heard the Antichrist **is coming**, even now many antichrists have come.

100. Translate the following sentences into Greek. The answers are provided.

1) I am a sinner.
2) He is being saved.

3) I am being saved by him.
4) I am not being loosed.
5) He saves me.
1) **εἰμι ἁμάρτωλος.**
2) **σώζεται.**
3) **σώζομαι ὑπ' αὐτοῦ.**
4) **οὐ λύομαι.**
5) **σώζει με.**
Now translate the Greek sentences above into English.

101. **Recognition Drill**

a) I am being saved by him./b) She is being saved by him./c) He saves me./d) The sinner enters a house./e) They enter into the synagogue./f) We are being saved by him./g) I answer the voice./h) We see the synagogue./i) Ye are being loosed by him./j) I hear the voice.

Match the Greek with the translation. Answers are provided below.

_____ **1. ἀποκρίνομαι τῇ φωνῇ.**
_____ **2. σώζομεθα ὑπ' αὐτοῦ.**
_____ **3. λύεσθε ὑπ' αὐτοῦ.**
_____ **4. βλέπομεν τὴν συναγωγήν.**
_____ **5. ἀκούω τῆς φωνῆς.**
_____ **6. εἰσπορεύονται εἰς τὴν συναγωγήν.**
_____ **7. ὁ ἁμάρτωλος εἰσέρχεται οἶκον.**
_____ **8. σώζει με.**
_____ **9. σώζεται ὑπ' αὐτοῦ.**
_____ **10. σώζομαι ὑπ' αὐτοῦ.**

Answers: 1g, 2f, 3i, 4h, 5j, 6e, 7d, 8c, 9b, 10a

NINETEEN

Καὶ ἐγγὺς ἦν τὸ πάσχα τῶν Ἰουδαίων.
And the Passover of the Jews was near.
John 2:13a

Imperfect Active Indicative

102. Vocabulary

αἴρω I take up, I take away

ἀναβαίνω I go up (**ἀνα** means "up")

ἀποθνήσκω I die

ἀποκτείνω I kill

ἀποστέλλω I send (with a commission) (apostle)

ἄρτος, ὁ bread, piece of bread, loaf of bread

βαίνω I go; Verb does not appear in GNT, but combined with other elements to form other verbs.

ἐσθίω I eat

κατά (prep) with the genitive, "against," with the accusative "according to"

καταβαίνω I go down

οὐκέτι no longer (adverb)

παραλαμβάνω I receive, I take along

πέμπω I send (in a general sense)

τότε then (adverb)

σύν (prep) with dative, "with." Similar to **μετά** with the genitive

παρά (prep) with genitive "from;" with dative "beside;" with accusative "alongside of" (parallel)

μέν ... δέ "on the one hand ... on the other hand." Or just translate as "but."

103. In Greek, in the present tense, there is no special form of the verb to indicate continued action. Therefore, **λύω** can mean either "I loose" or "I am loosing" (continued action). We see that a definite distinction is made in English between continued action in the present tense, but not so much in Greek. "I loose" would be the simple present tense and "I am loosing" would indicate continued action in the present tense. This is sometimes called "the present continuous."

However, in the past tense, a distinction in Greek is made. The tense which indicates simple past tense in Greek is called the aorist. It will be studied later. The tense which is used to express continued action in the past tense is called the imperfect.

104. The aorist active indicative of **λύω** means "I loosed," etc. The imperfect active indicative of **λύω** indicates continued action in the past: "I was loosing," etc. The student is encour-

aged to carefully note the difference. The forms of the imperfect active indicative of **λύω** are given below.

SINGULAR

1P **ἔλυον** I was loosing
2P **ἔλυες** you were loosing
3P **ἔλυε (ν)** he/she/it was loosing

PLURAL

1P **ἐλύομεν** we were loosing
2P **ἐλύετε** you (ye) were loosing
3P **ἔλυον** they were loosing

THE IMPERFECT, like the indicative of the other secondary tenses, has an augment (**ε**) at the beginning of the stem of the verb. This augment indicates past time. (The primary tenses are the present [I loose/am loosing], the future [I will loose] and the perfect [I have loosed]; the secondary tenses are the imperfect [I was loosing], the aorist [I loosed], and the pluperfect [I had loosed].) These concepts are discussed in detail in Chapter Twenty-Seven.

Note that the personal endings in the active of the secondary tenses, which includes the imperfect, are: -**ν**, -**ς**, none, **-μεν**, -**τε**, -**ν** (or -**σαν**)

105. **Recognition Tip**

In the active tenses, not in the middle and passive tenses, the **-μεν** ending ("we" – first person plural) and the -**τε** ending ("you" or "ye" – second person plural) are consistent throughout the various conjugations.

In addition, it will be observed that the verb form of the first person singular and the third person plural are the same. Only by looking at the context of a sentence will the student be able to determine whether **ἔλυον** means "I was loosing" or "they were loosing."

THE VARIABLE VOWEL is placed between the stem and the personal endings. In the imperfect, as in the present tense, **ο**, comes before **μ** and **ν**, and **ε** before other letters.

THE AUGMENT in verbs that begin with a consonant consists of an **ε** prefixed to the stem. In the case of **λύω**, the stem is **λυ-**.Thus **ἔλυον** means "I was loosing" and **ἐγίνωσκον** means "I was knowing."

WITH VERBS BEGINNING WITH A VOWEL, the vowel is "lengthened" to form the augment:

α is lengthened to **η**
ε is lengthened to **η**
ι remains the same
ο is lengthened to **ω**
υ remains the same

106. **Augment of Compound Verbs**

Compound verbs are those to which a preposition has been affixed. In the case of these verbs the augment is placed after the preposition and before the stem. If the preposition ends in a vowel, that vowel is usually dropped not only before a verb which begins with a vowel, but before the augment is inserted. Note the examples below:

The imperfect form of **ἐκβαλλω** ("I cast out.") is **ἐξέβαλλω** ("I was casting out."). Notice that the augment **ε** is placed after the preposition **ἐκ**. Before a vowel, **ἐκ** changes to **ἐξ** to facilitate pronunciation.

The imperfect of **ἀποκτείνω** ("I kill.") is **ἀπέκτεινον** ("I was killing."). Notice that the final vowel of the preposition, **ἀπό**, is dropped and the augment **ε** is put in its place. The imperfect of **ἀπάγω** ("I go away.") is **ἀπῆγον** ("I was going away"). In this case the final vowel of the preposition, **ἀπό**, is dropped. Then the **α** of **ἄγω** ("I lead") is lengthened to **η**, as the chart at the end of section 105 indicates.

107. **Greek Verses Featuring Imperfect Verbs**

Imperfect verbs, as well as the English equivalents, are bolded and underlined.

MARK 5:8

<u>ἔλεγεν</u> γὰρ αὐτῷ, Ἔξελθε τὸ πνεῦμα τὸ ἀκάθαρτον ἐκ τοῦ ἀνθρώπου.

For **<u>he was saying</u>** to him, "Come out of the man, unclean spirit."

MARK 4:26

Καὶ **<u>ἔλεγεν</u>**, Οὕτως ἐστὶν ἡ βασιλεία τοῦ θεοῦ ὡς ἄνθρωπος βάλῃ τὸν σπόρον ἐπὶ τῆς γῆς.

And **<u>he was saying</u>**, "Thus is the kingdom of

God, as a man casting the seed upon the earth."

JOHN 4:33

ἔλεγον οὖν οἱ μαθηταὶ πρὸς ἀλλήλους μή τις ἤνεγκεν αὐτῷ φαγεῖν;

Therefore the disciples **were saying** to one another, "Has no one brought him (anything) to eat?"

JOHN 9: 9 – 10

ἄλλοι **ἔλεγον** ὅτι οὗτός ἐστιν· ἄλλοι δὲ ὅτι ὅμοιος αὐτῷ ἐστιν. ἐκεῖνος **ἔλεγεν** ὅτι ἐγώ εἰμι. **ἔλεγον** οὖν αὐτῷ·πῶς ἀνεῴχθησάν σου οἱ ὀφθαλμοί;

Others **were saying**, "He's the one." (Still) others were saying, "No, he resembles him." (But) that man **was saying**, "I am he." They **were saying** therefore to him, "How then were your eyes opened?"

MARK 4:34A

χωρὶς δὲ παραβολῆς **οὐκ ἐλάλει** αὐτοῖς.

But apart from parable(s) **he was not speaking** to them.

MARK 2:13

Καὶ ἐξῆλθεν πάλιν παρὰ τὴν θάλασσαν· καὶ πᾶς ὁ ὄχλος **ἤρχετο** πρὸς αὐτόν, καὶ

ἐδίδασκεν αὐτούς.
And he went out again by the lake. And the entire crowd **was coming** to him, and he **was teaching** them.

MARK 6:15

ἄλλοι δὲ **ἔλεγον** ὅτι Ἠλίας ἐστίν. ἄλλοι δὲ **ἔλεγον** ὅτι προφήτης ὡς εἷς τῶν προφητῶν.
And others **were saying** that he is Elijah. And others **were saying** that (he is) a prophet like one of the prophets.

MARK 4:41

Καὶ ἐφοβήθησαν φόβον μέγαν, καὶ **ἔλεγον** πρὸς ἀλλήλους, Τίς ἄρα οὗτός ἐστιν ὅτι καὶ ὁ ἄνεμος καὶ ἡ θάλασσα ὑπακούει αὐτῷ;
And they were sore afraid, and **were saying** to one another, "Who, then, is this man, that even the wind and the sea obey him?"

MARK 4:24A

Καὶ **ἔλεγεν** αὐτοῖς, Βλέπετε τί ἀκούετε.
And **he was saying** to them, "Beware what you hear."

MARK 3:23

Καὶ προσκαλεσάμενος αὐτοὺς ἐν παραβολαῖς **ἔλεγεν** αὐτοῖς, Πῶς δύναται Σατανᾶς Σατανᾶν ἐκβάλλειν;

And he called them to himself (and) **was speaking** to them in parables, "how is Satan able to cast out Satan?"

MARK 4:2

καὶ **ἐδίδασκεν** αὐτοὺς ἐν παραβολαῖς πολλά, καὶ **ἔλεγεν** αὐτοῖς ἐν τῇ διδαχῇ αὐτοῦ.

And he **was teaching** them using many parables, and **was speaking** to them in his teaching.

MARK 11:14

καὶ ἀποκριθεὶς εἶπεν αὐτῇ μηκέτι ἐκ σοῦ εἰς τὸν αἰῶνα μηδεὶς καρπὸν φάγοι. καὶ **ἤκουον** οἱ μαθηταὶ αὐτοῦ.

And he answered and said to it. "Let no one eat fruit from you ever again." And his disciples **were listening.**

108. Translate the following into Greek using the imperfect. Answers are provided. Clues to the answers are contained in the verses quoted above in section 107.

1) His disciples were listening.

2) Therefore the disciples were saying (speaking) to one another.

3) And he was teaching them in many parables.

1) οἱ μαθηταὶ αὐτοῦ ἤκουον.

2) οὖν οἱ μαθηταὶ ἔλεγον πρὸς ἀλλήλους.

3) καὶ ἐδίδασκεν αὐτοὺς ἐν παραβολαῖς πολλά.

. . .

109. **Recognition Drill**

Match the Greek with the translation. Answers are provided below.

a) His disciples were listening./b) The sinner was speaking to him./c) They were speaking the words to men./d) He was loosing me./e) I answer the voice./f) We are being saved by him./g) I am in the good temple./h) The disciples were speaking./i) I was casting out a sinner./j) They were loosing a sinner.

_____ **1. ἀποκρίνομαι τῇ φωνῇ.**

_____ **2. σῴζομεθα ὑπ' αὐτοῦ.**

_____ **3. εἰμὶ ἐν τῷ ἀγαθῷ ἱερῷ.**

_____ **4. ἔλεγον τοὺς λόγους ἀνθρώποις.**

_____ **5. ἔλυον ἁμάρτωλον.**

_____ **6. ἐξέβαλλω ἁμάρτωλον.**

_____ **7. ὁ ἁμάρτωλος ἔλεγον πρὸς αὐτόν.**

_____ **8. ἔλυεν με.**

_____ **9. οἱ μαθηταὶ ἔλεγον.**

_____ **10. οἱ μαθηταὶ αὐτοῦ ἤκουον.**

Answers: 1e, 2f, 3g, 4c, 5j, 6i, 7b, 8d, 9h, 10a

TWENTY

τὸ φῶς ἐλήλυθεν εἰς τὸν κόσμον καὶ ἠγάπησαν οἱ ἄνθρωποι μᾶλλον τὸ σκότος ἢ τὸ φῶς.
The Light came into the world and men loved the darkness more than the light.
John 3:19a

110. **Imperfect Active Indicative of εἰμί ("I am")**

This verb form is called the imperfect in Greek grammar, but the student will recognize the English forms as the simple past tense of the verb "to be."

SINGULAR

1P **ἤμην** I was
2P **ἦς** you were
3P **ἦν** he (she or it) was

. . .

PLURAL

1P **ἦμεν** we were
2P **ἦτε** you were
3P **ἦσαν** they were

NOTICE that in the imperfect the **εἰ** of the present active indicative of **εἰμί** has augmented and lengthened to **η** and that the following personal endings in the active of the secondary tenses, which we have encountered previously, have been added: -**ν**, -**ς**, none, -**μεν**, -**τε**, -**ν** (or **-σαν**). Note that the endings -**μεν**, and -**τε** can be reliably counted on to be the forms of the first-person plural (we) and the second person plural (you [ye]) respectively.

111. **Recognition Tip**

For purposes of comparison with the imperfect of **εἰμί**, above, the conjugation of the present tense of **εἰμί**, which has already been studied, is repeated below. The student will note the present tense and the imperfect have some similarities.

Look at the 1P and the 2P plural forms of both the imperfect tense and the present tense and note that the verb endings (the **μεν** and the **τε**) are the same in both cases. These types of similarities serve to aid in recognition.

SINGULAR

1P **εἰμί** I am
2P **εἶ** you are
3P **ἐστί(ν)** he (she or it) is

. . .

PLURAL

1P **ἐσμέν** we are
2P **ἐστέ** you are
3P **εἰσί(ν)** they are

112. **Greek Verses with the imperfect of εἰμί**

The imperfect of **εἰμί**, as well as the English equivalents, are bolded and underlined.

MARK 2:18

Καὶ **<u>ἦσαν</u>** οἱ μαθηταὶ Ἰωάννου καὶ οἱ τῶν Φαρισαίων νηστεύοντες. καὶ ἔρχονται καὶ λέγουσιν αὐτῷ· διατὶ οἱ μαθηταὶ Ἰωάννου καὶ οἱ τῶν Φαρισαίων νηστεύουσιν, οἱ δὲ σοὶ μαθηταὶ οὐ νηστεύουσι;

And the disciples of John and the Pharisees **<u>were</u>** fasting. And they come and say to him, "Why is it that the disciples of John and of the Pharisees fast, and your disciples do not fast?"

LUKE 2:8A

Καὶ ποιμένες **<u>ἦσαν</u>** ἐν τῇ χώρᾳ.

And **<u>there were</u>** shepherds in the country.

MARK 5:42

Καὶ εὐθὺς ἀνέστη τὸ κοράσιον καὶ περιεπάτει, **<u>ἦν</u>** γὰρ ἐτῶν δώδεκα.

And the young girl immediately rose up and

walked, for she **was** twelve years (old).

MARK 11:30

τὸ βάπτισμα τὸ Ἰωάννου ἐξ οὐρανοῦ **ἦν** ἢ ἐξ ἀνθρώπων;

Was the baptism of John from heaven or from men?

JOHN 3:1 – 2a

Ἦν δὲ ἄνθρωπος ἐκ τῶν Φαρισαίων, Νικόδημος ὄνομα αὐτῷ, ἄρχων τῶν Ἰουδαίων.

And there **was** a man of the Pharisees by the name of Nicodemus, a leader of the Jews.

MARK 1:23

Καὶ εὐθὺς **ἦν** ἐν τῇ συναγωγῇ αὐτῶν ἄνθρωπος ἐν πνεύματι ἀκαθάρτῳ.

And immediately there **was** in their synagogue a man with an unclean spirit.

MARK 15:39B

Ἀληθῶς οὗτος ὁ ἄνθρωπος υἱὸς θεοῦ **ἦν**.

Truly this man **was** the Son of God!

JOHN 1: 1–2

Ἐν ἀρχῇ **ἦν** ὁ Λόγος, καὶ ὁ Λόγος **ἦν** πρὸς τὸν Θεόν, καὶ Θεὸς **ἦν** ὁ Λόγος. Οὗτος **ἦν** ἐν ἀρχῇ πρὸς τὸν Θεόν.

In the beginning **was** the Word, and the Word **was** with God and the Word **was** God. He **was** in the beginning with God.

JOHN 1: 44

ἦν δὲ ὁ Φίλιππος ἀπὸ Βηθσαϊδά, ἐκ τῆς πόλεως ᾿Ανδρέου καὶ Πέτρου.

And Philip **was** from Bethsaida, from the city of Andrew and Peter,

TRANSLATION PRACTICE with the imperfect of **εἰμί**

Translate the following English sentences into Greek and then translate the Greek sentences into English. The answers are given below. You will find clues in the Greek verses quoted above.

1) We were with God,

2) And God was the Word.

3) Philip was from the city.

4) And the disciples were not with God in the beginning.

1) **ἦμεν πρὸς τὸν Θεόν.**

2) **καὶ Θεὸς ἦν ὁ Λόγος.**

3) **Φίλιππος ἦν ἐκ τῆς πόλεως.**

4) **Καὶ οἱ μαθηταὶ οὐκ ἦσαν πρὸς τὸν Θεὸν ἐν ἀρχῇ.**

113. **Recognition Drill**

Match the Greek with the translation. Answers are provided below.

a) We were with God./b) The disciples were not with

God./c) She was from the city./d) In the beginning was the Word./e) We were in the good temple./f) Ye were not with God in the beginning./g) You were not in heaven./h) I was in our synagogue./i) They were in their temple./j) We were with the brother.

_____ **1. οὐκ ἦτε ἐν ἀρχῇ πρὸς τὸν Θεὸν.**
_____ **2. οὐκ ἦς ἐν τῷ οὐρανῷ.**
_____ **3. ἦμεν ἐν τῷ ἀγαθῷ ἱερῷ.**
_____ **4. ἤμην ἐν τῇ συναγωγῇ ἡμῶν.**
_____ **5. ἦσαν ἐν τῷ ἱερῷ αὐτῶν.**
_____ **6. ἦμεν σὺν τῷ ἀδέλφῳ.**
_____ **7. ἐν ἀρχῇ ἦν ὁ Λόγος.**
_____ **8. ἦν ἐκ τῆς πόλεως.**
_____ **9. ἦμεν πρὸς τὸν Θεόν.**
_____ **10. οἱ μαθηταὶ οὐκ ἦσαν πρὸς τὸν Θεὸν.**

Answers: 1f, 2g, 3e, 4h, 5i, 6j, 7d, 8c, 9a, 10b

TWENTY-ONE

Ὁ Πατὴρ ἀγαπᾷ τὸν Υἱόν καὶ πάντα δέδωκεν ἐν τῇ χειρὶ αὐτοῦ.
The Father loves the Son and has given all things into His hands.
John 3:35

Imperfect Passive Indicative

114. Vocabulary

ἀπέρχομαι I go away, (deponent verb)
βιβλίον, τό book, (bibliography)
δαιμόνιον, τό demon
δέχομαι I receive, welcome (deponent verb)
ἐκπορεύομαι I go out (deponent verb)
ἔργον, τό work (noun)
ἔτι still, yet (adverb)
θάλασσα, ἡ lake
καί and, also, even (conjunction); **καί ... καί**, both ... and

κατέρχομαι I go down (deponent verb)

οὐδέ and not, nor, not even (conjunction); **οὐδέ ... οὐδέ**, neither ... nor

οὔπω (adverb) not yet (adverb)

περί (prep) concerning, about (genitive); around (accusative)

πλοῖον, τό boat

συένρχομαι I come together (deponent verb)

ὑπέρ (prep) in behalf of (genitive); above (accusative) (hyperbole)

115. The imperfect passive indicative of **λύω** is as follows:

SINGULAR

1P **ἐλυόμην** I was being loosed
2P **ἐλύου** you were being loosed
3P **ἐλύετο** he was being loosed

PLURAL

1P **ἐλυόμεθα** we were being loosed
2P **ἐλύεσθε** you were being loosed
3P **ἐλύοντο** they were being loosed

116. The imperfect tense is a past tense and therefore adds an augment to the stem of the verb. In this case the augment is **ε**.

The imperfect is a secondary tense. The personal endings in the passive of the secondary tenses are: **-μην**, **-σο**, **-το**, **-μεθα**, **-σθε**, **-ντο**. The variable vowel comes into play as it does in the imperfect active and all three voices (active, middle and passive) of the present tense: **ο** before **μ** and **ν** and **ε** before other letters.

The second person singular, **ἐλύου**, is a contracted form.

117. Once again, for a detailed discussion of the attributes of primary and secondary tenses, as well as their verb endings, see Chapter Twenty-Seven.

118. **The Neuter Plural Subject**

A neuter subject may have a verb in the singular. For example:

τὰ δαιμόνια ἀπέρχεται. This sentence means, "The demons go away."

119. **Other uses of καί**

The meaning of **καί**, translated as "and" has already been mentioned. **καί** can also mean "also" or "even." When used in these ways, it comes before the word with which it is logically connected, For example:

τοῦτο δὲ καὶ λέγει. = "But this he also says."

βλέπουσι καὶ τὰ δαιμόνια τὸ ἔργον. = "Even the demons see the work."

TWENTY-TWO

Ἔρχεται γυνὴ ἐκ τῆς Σαμαρείας ἀντλῆσαι ὕδωρ.
A woman from Samaria comes to draw water.
John 4:7a

Future Active and Future Middle Indicative

120. The vocabulary below contains a number of new verbs as well as their future tense forms. Notice that some of the verbs have future deponent forms. That is, the form looks like a middle form, and is designated as "deponent future," but the meaning is simple future.

121. Vocabulary

ἀναβλέψω I shall look up/shall receive my sight (future of **ἀναβλέπω** [I look up/receive my sight])

βήσομαι I shall go (deponent future of **βαίνω** [I go])

βλέψω I shall see (future of **βλέπω** [I see])

γενήσομαι I shall become (deponent future of **γίνομαι** [I become, I am])

γνώσομαι I shall know (deponent future of **γινώσκω** [I know, am knowing, do know])

διδάξω I shall teach (future of **διδάσκω** [I teach, am teaching, do teach])

διώξω I shall pursue/persecute (future of **διώκω** [I pursue/persecute])

δοξάσω I shall glorify (future of **δοξάζω** [I glorify])

ἐλεύσομαι I shall come/go (deponent future of **ἔρχομαι** [I come, go])

ἕξω I shall have (future of **ἔχω** [I have])

κηρύξω I shall proclaim/preach (future of **κηρύσσω** [I proclaim/preach])

λήμψομαι I shall take (deponent future of **λαμβάνω** [I take, am taking, do take])

προσεύξομαι I shall pray (deponent future of **προσεύχομαι** [I pray, am praying, do pray])

τυφλός, ὁ blind man

122. It is important to note that some of the verbs listed in the above vocabulary are deponent in the future tense, but not in the present tense.

The future active indicative of **λύω** is provided below. The verb endings should be familiar because they have already been encountered in the present active indicative.

SINGULAR

1P **λύσω** I shall loose

2P **λύσεις** you shall loose

3P **λύσει** he/she/it shall loose

. . .

PLURAL

1P **λύσομεν** we shall loose
2P **λύσετε** you (ye) shall loose
3P **λύσουσι (ν)** they shall loose

123. The future middle indicative of **λύω** is provided below. Once again, the verb endings should be familiar.

SINGULAR

1P **λύσομαι** I shall loose for myself
2P **λύσῃ** you shall loose for yourself
3P **λύσεται** he/she/it shall loose for himself

PLURAL

1P **λυσόμεθα** we shall loose for ourselves
2P **λύεσθε** you (ye) shall for yourselves)
3P **λύσονται** they shall for themself

124. **Recognition Tips**

IN THIS LESSON the future active indicative and the future middle indicative are introduced. The student is encouraged to keep two thoughts in mind:

The future active indicative is identical to the present active indicative, except that in most cases a **σ** is added to the stem before the personal endings are attached. The future forms given in the vocabulary list above, which do not form the future by adding the **σ**, are irregular future verb forms.

The future middle indicative is provided so that the student can see how one arrives at the forms for those verbs which are deponent in the future. The future middle indicative forms ("I shall loose for myself.") is rare. Normally this middle form is used for the deponent future, which means that the English translation for a future middle form will be, for example, **ἐλεύσομαι** ("I shall come."), which is active in meaning.

THE STUDENT WILL RECALL that all three voices of the present and imperfect tenses are formed using the present stem. To this stem the personal endings are added once the variable vowel (**ο** or **ε**) has been added to the stem.

In the case of the future active and middle indicative, the personal endings are added to what is called the future stem, which, with most verbs, is the present stem with the addition, as mentioned above, of a **σ**. Therefore, in the case of **λύω**, **λυ-** is the is the present tense stem and **λυσ-** is the future stem.

The student will also recall that the future is a primary tense. Therefore, the future, like the present tense, will receive the primary personal endings. This means that the future active and middle indicative are conjugated just like the present active and middle, except that the future has **λυσ-** as the stem and the present has **λυ-**.

125. **Changes in the stem in the future active and middle indicative**

Please keep in mind that our goal is to <u>recognize</u> the text, not to possess a complete knowledge of every grammatical rule in Greek. With this thought as a backdrop, let us consider a few aberrations of the future tense stem.

WHEN THE GREEK verb stem ends in a consonant, the addition of the tense suffix **σ** brings two consonants together. This meeting of consonants results in a few changes in the stem.

When **π**, **β**, or **φ** meet and combine with **σ** the result is the double consonant **ψ**. **π**, **β**, and **φ** are called labial mutes because they are pronounced using the lips.

Examples: The future of **γράφω** is **γράψω**; the future of **πέμπω** is **πέμψω**.

When **κ**, **γ**, or **χ** meet and combine with **σ**, the result is the double consonant **ξ**. **κ**, **γ**, and **χ** are called palatal mutes because they are pronounced using the palate.

Examples: The future of **ἄρχω** ("I rule") is **ἄρξω**; the future of **ἄγω** ("I lead") is **ἄξω**.

τ, **δ**, and **θ** drop out before the **σ** and the **σ** replaces them. **τ**, **δ**, and **θ** are called lingual mutes because they are pronounced using the tongue.

Example: The future of **πείθω** (I persuade) is **πείσω**.

The future tense of Greek verbs cannot always be determined based on rules. In many cases the student will have to consult a lexicon to determine the future form of any given verb.

126. **Future Active Indicative of εἰμί**

THE FUTURE ACTIVE indicative of **εἰμί** is provided below.

Notice that even though it is the future active tense, it has middle verb endings.

SINGULAR

1P **ἔσομαι** I shall be
2P **ἔσῃ** you shall be
3P **ἔσται** he/she/it shall be

PLURAL

1P **ἐσόμεθα** we shall be
2P **ἔσεσθε** you (ye) shall be
3P **ἔσονται** they shall be

127. **Greek Verses in the Future Tense**

Occurrences of the Greek future, as well as the English equivalents, are bolded and underlined.

LUKE 1:14A

Καὶ **<u>ἔσται</u>** χαρά σοι καὶ ἀγαλλίασις.
And (there) **<u>will be</u>** joy for you and gladness.

LUKE 1:33

καὶ **<u>βασιλεύσει</u>** ἐπὶ τὸν οἶκον Ἰακὼβ εἰς τοὺς αἰῶνας, καὶ τῆς βασιλείας αὐτοῦ οὐκ **<u>ἔσται</u>** τέλος.
And **<u>he will reign</u>** over the house of Jacob forever. There **<u>will be</u>** no end to his Kingdom.

. . .

MARK 11:29A

ὁ δὲ Ἰησοῦς ἀποκριθεὶς εἶπεν αὐτοῖς,
<u>ἐπερωτήσω</u> ὑμᾶς κἀγὼ ἕνα λόγον.
And Jesus answered (and) said to them,
"I, also, **<u>will ask</u>** you a question."

MATTHEW 1:21

<u>τέξεται</u> δὲ υἱὸν καὶ **<u>καλέσεις</u>** τὸ ὄνομα
αὐτοῦ Ἰησοῦν· αὐτὸς γὰρ **<u>σώσει</u>** τὸν λαὸν
αὐτοῦ ἀπὸ τῶν ἁμαρτιῶν αὐτῶν.
<u>She shall</u> bear a son and **<u>you shall call</u>** his
name Jesus, for **<u>he shall save</u>** his people from
their sins.

MATTHEW 3:11B

αὐτὸς ὑμᾶς **<u>βαπτίσει</u>** ἐν Πνεύματι Ἁγίῳ.
He shall baptize you in the Holy Spirit.

(Notice that the preposition **ἐν** takes the dative case and that the two Greek nouns after it are in the dative singular and end in **ι**, which is the sign of the dative singular. As mentioned before, looking for the final **ι** as a sign of the dative singular works most of the time. An exception would be οἱ ἄνθρωποι, where both words end in **ι**, but are in the nominative plural.)

JOHN 14:26

ὁ δὲ παράκλητος, τὸ Πνεῦμα τὸ Ἅγιον ὃ
<u>πέμψει</u> ὁ πατὴρ ἐν τῷ ὀνόματί μου,
ἐκεῖνος ὑμᾶς **<u>διδάξει</u>** πάντα καὶ

ὑπομνήσει ὑμᾶς πάντα ἃ εἶπον ὑμῖν.
And the Comforter, the Holy Spirit, Whom the father **shall send** in my name, that one **shall teach** you all things and **will remind** you of all the things that I said to you.

MARK 1:8

ἐγὼ μὲν ἐβάπτισα ὑμᾶς ἐν ὕδατι, αὐτὸς δὲ **βαπτίσει** ὑμᾶς ἐν Πνεύματι Ἁγίῳ.
I baptized you in water, but **he shall baptize** you in (the) Holy Spirit.

LUKE 1:32

οὗτος **ἔσται** μέγας καὶ υἱὸς ὑψίστου **κληθήσεται**, καὶ **δώσει** αὐτῷ κύριος ὁ θεὸς τὸν θρόνον Δαυὶδ τοῦ πατρὸς αὐτοῦ.
He **will be** great and **will be called** the Son of the Most High. The Lord God **will give** him the throne of his father David.

MARK 13:8

ἐγερθήσεται γὰρ ἔθνος ἐπ' ἔθνος καὶ βασιλεία ἐπὶ βασιλείαν, **ἔσονται** σεισμοὶ κατὰ τόπους, **ἔσονται** λιμοί· ἀρχὴ ὠδίνων ταῦτα.
For nation **will rise** against nation, and kingdom against kingdom. There **will be** earthquakes in various places. There **will be** famines and troubles. These things are the beginning of birth pains.

. . .

MARK 13:19A

<u>ἔσονται</u> γὰρ αἱ ἡμέραι ἐκεῖναι θλῖψις.
For in those days there **<u>will be</u>** oppression.

MARK 13:13

καὶ **<u>ἔσεσθε</u>** μισούμενοι ὑπὸ πάντων διὰ τὸ ὄνομά μου. ὁ δὲ ὑπομείνας εἰς τέλος οὗτος σωθήσεται.
And **<u>you will be</u>** hated by all men for my name's sake, but he who endures to the end will be saved.

LUKE 1:34

εἶπεν δὲ Μαριὰμ πρὸς τὸν ἄγγελον, Πῶς **<u>ἔσται</u>** τοῦτο, ἐπεὶ ἄνδρα οὐ γινώσκω;
Mary said to the angel, "How **<u>shall</u>** this **<u>be</u>**, seeing I am a virgin?"

128. **Translation Exercises for the Future Active Indicative**

Translate the following English sentences into Greek. After that cover the English sentences and translate the Greek into English. Answers are provided below.

1) I shall see my brother and shall glorify my good God.

2) He shall persecute your brethren.

3) We shall proclaim the Word of God.

1) **βλέψω τὸν ἀδελφόν μου καὶ δοξάσω τὸν ἀγαθὸν θεόν μου.**

2) **διώξει τοὺς ἀδέλφους σου.**
3) **κηρύξομεν τὸν λόγον θεοῦ.**

129. **Recognition Drill**

Match the Greek with the translation. Answers are provided below.

a) I shall see my brother./b) I shall glorify my good God./c) Ye shall be with the brother./d) They shall be from the city./e) In the beginning was the Word./f) We shall be in their temple./g) You shall be in our synagogue./h) He shall be in the good temple./i) He shall persecute your brethern./j) We shall proclaim the Word of God.

_____ **1. κηρύξομεν τὸν λόγον θεοῦ.**
_____ **2. διώξει τοὺς ἀδέλφους σου.**
_____ **3. ἔσται ἐν τῷ ἀγαθῷ ἱερῷ.**
_____ **4. ἔσῃ ἐν τῇ συναγωγῇ ἡμῶν.**
_____ **5. ἐσόμεθα ἐν τῷ ἱερῷ αὐτῶν.**
_____ **6. ἔσεσθε σύν τῷ ἀδέλφῳ.**
_____ **7. ἐν ἀρχῇ ἦν ὁ Λόγος.**
_____ **8. ἔσονται ἐκ τῆς πόλεως.**
_____ **9. δοξάσω τὸν ἀγαθὸν θεόν μου.**
_____ **10. βλέψω τὸν ἀδελφόν μου.**

Answers: 1j, 2i, 3h, 4g, 5f, 6c, 7e, 8d, 9b, 10a

TWENTY-THREE

Πνεῦμα ὁ Θεός, καὶ τοὺς προσκυνοῦντας αὐτὸν ἐν
πνεύματι καὶ ἀληθείᾳ δεῖ προσκυνεῖν.
God is a Spirit, and it is necessary for those who worship Him
to worship Him in Spirit and truth.
John 4:24

First Aorist Active and Middle Indicative

130. Vocabulary

ἀπολύω, ἀπολύσω, **ἀπέλυσω** I release, I shall release, I released (**ἀπολύω** can also mean "I destroy" or "I dismiss")

ἐκρήυξα I preached, I proclaimed; (first aorist of **κρηύσσω**)

ἐπιστρέφω, **ἐπιστρέψω**, **ἐπέστρεψα** I turn (return), I shall turn, I turned

ἤδη already (adverb)

θαυμάζω, θαυμάσω, ἐθαύμασα I marvel, I shall marvel, I marveled

θεραπεύω, θεραπεύσω, ἐθεράπευσα I heal, I shall heal, I healed (therapeutic)

πείθω, πείσω, ἔπεισα I persuade, I shall persuade, I persuaded

πιστεύω, πιστεύσω, ἐπίστευσα I believe, I shall believe, I believed

ὑποστρέφω, ὑποστρέψω, ὑπέστερψα I return, I shall return, I returned

THE VERB FORMS listed above that end with **σα**, **ζα**, and **ψα** are first aorist (simple past) forms.

There are two aorist tenses in Greek, the first aorist and the second aorist. The second aorist will be studied in the next lesson. The first aorist is not a different tense from the second aorist. The two terms simply represent two different ways of forming the same tense (simple past) of a verb. Certain Greek verbs form their aorist in one manner (the first aorist); others form their aorist in another manner (second aorist), which involves a vowel change in the body of the verb. English has a similar concept. The English "I believed" is not a different tense from "I sang," but the two verbs form the simple past in two distinct and different ways. The verb "believe" forms its simple past by adding a "d" to the infinitive (dictionary) form, resulting in "I believed, "and the verb "sing" forms its simple past by means of a vowel change, that is, the "i" of the infinitive changes to "a" to form the simple past form, "I sang."

. . .

AS WITH THE IMPERFECT, the aorist refers to past time. However, the imperfect is used to describe continuous action in the past ("I was loosing"), whereas the aorist refers to a single occurrence of action in the past (I loosed).

Sometimes, for clarity, English translations render the Greek aorist by the English perfect tense. Therefore **ἔλυσα** (first aorist tense) may mean "I have loosed" (perfect tense) or "I loosed" (aorist). For example:

ἔλυσα τὸν δοῦλον μου. "I have loosed my servant." In this case, although **ἔλυσα** is aorist and means "I loosed," the action of the verb has taken place at some unspecified time in the past.

τότε ἔλυσα τὸν δοῦλον μου. "Then I loosed my servant." Here, the word "then,"(**τότε**) has been added, which definitely fixes the time of the past action. However, the aorist is rather fluid in nature and "I have loosed" and "I loosed" are both possible in each of the above examples.

131. Below is the first aorist indicative of **λύω**.

SINGULAR

1P **ἔλυσα** I loosed
2P **ἐλύσας** you loosed
3P **ἔλυσε(ν)** he/she/it loosed

PLURAL

1P **ἐλύσαμεν** we loosed
2P **ἐλύσατε** you (ye) loosed
3P **ἔλυσαν** they loosed

. . .

NOTE THAT THE AORIST, like the imperfect, is a secondary tense, and therefore has the augment (**ε**). The augment is the same for the aorist as it is for the imperfect. The aorist, like the imperfect, also has the secondary endings. For review, the personal endings in the active of the secondary tenses are: **-ν**, **-ς**, none, **-μεν**, **-τε**, **-ν** (or **-σαν**)

NOTICE that the **ν** is dropped in the first-person singular of the aorist. The aorist does not employ a variable vowel, as the imperfect and the future do. In the aorist, the tense suffix **σα** is added to the verb stem. Therefore, where the future has **σο/ε**, the aorist employs **σα**.

Also notice that in the third person singular the **σα** is replaced by **σε**. **ἔλυσε(ν)** (he loosed) can have the movable **ν**, like **ἔλυε (ν)** (he was loosing) of the imperfect.

132. **Recognition Tips**

The student should take note of the make-up of the various forms. For example, **ἐλύσαμεν** is composed of **ἐ/λύ/σα/μεν**. **ἐ** is the augment, **λύ** is the stem of the verb, **σα** is the sign of the first aorist, and **μεν** is the secondary personal ending of the first person plural active.

As mentioned above, **σα** and **σε** after the verb stem signal the appearance of the first aorist. Remember that since the first aorist is a past tense, it will have an augment at the beginning of the verb. Most of the time the augment will be **ε**, as discussed, but that augment can lengthen.

133. Listed below is the first aorist middle indicative of **λύω**.

. . .

SINGULAR

1P **ἐλυσάμην** I loosed for myself
2P **ἐλύσω** you loosed for yourself
3P **ἐλύσατο** he/she/it loosed for himself

PLURAL

1P **ἐλυσάμεθα** we loosed for ourselves
2P **ἐλύσασθε** you (ye) loosed for yourselves
3P **ἐλύσαντο** they loosed for themselves

THE FORMS of the first aorist middle indicative where **ἐλυσάμην** is translated "I loosed for myself," etc., are rare. These forms are primarily used for verbs that are deponent in the aorist, which means they would use the aorist middle indicative forms above, but the translation would be active. As an example, let's take the deponent verb **πορεύομαι**, which means "I go." Notice that it has middle/passive endings in the present tense, but the meaning is active. This verb is deponent in the future as well. **πορεύσομαι** means "I will go," not "I will go for myself." **πορεύομαι** is also deponent in the aorist and has the middle forms listed above. The form **ἐπορεύσαμήν**, which means simply "I went," is a first aorist form.

NOTE THAT THE AORIST MIDDLE, like the aorist active, has the secondary personal endings. For review purposes, the personal endings in the middle of the secondary tenses are: **-μην**, **-σο**, **-το**, **-μεθα**, **-σθε**, **-ντο**.

. . .

THE STUDENT SHOULD TAKE note of the make-up of the various forms. For example, **ἐλυσάμεθα** is composed of **ἐ/λυ/σά/μεθα**. **ἐ** is the augment, **λυ** is the stem of the verb, **σά** is the sign of the first aorist, and **μεθα** is the secondary personal ending of the first person plural middle.

BECAUSE THE VERB forms of the first aorist cannot be predicted with any certainly, the student, if in doubt, is advised to consult a Greek lexicon to determine the correct forms.

THE AORIST ACTIVE **Indicative versus the Imperfect Active Indicative**

134. Remember that the aorist active indicative expresses a completed event in the past and the imperfect active indicative expresses an ongoing event in the past.

For example, consider the English sentences below, in which the aorist and imperfect verb forms are bolded and underlined.

I **ate** a hearty meal yesterday. (Aorist – action completed. This tense is the "simple past." One could consider the action a **snapshot** of past action.)

She **was snacking** the entire afternoon. She **used to snack** the entire afternoon. (Imperfect. The action was ongoing and there is no indication that it was completed or "perfected." – thus the term "imperfect." One could consider the imperfect a **video** of past action.)

. . .

135. **Greek Verses in the First Aorist Tense**

Occurrences of the First Aorist, as well as the English equivalents, are bolded and underlined.

MARK 11:11a

Καὶ **<u>εἰσῆλθεν</u>** εἰς Ἱεροσόλυμα εἰς τὸ ἱερόν.
And **<u>he entered</u>** into Jerusalem (and) into the temple.

MARK 12:4A

Καὶ πάλιν **<u>ἀπέστειλεν</u>** πρὸς αὐτοὺς ἄλλον δοῦλον.
And again, **<u>he sent</u>** to them another servant.

MARK 8:27A

Καὶ **<u>ἐξῆλθεν</u>** ὁ Ἰησοῦς καὶ οἱ μαθηταὶ αὐτοῦ εἰς τὰς κώμας Καισαρείας τῆς Φιλίππου.
And Jesus **<u>went out</u>** with his disciples into the villages of Caesarea of Phillipi.

MARK 14:16

Καὶ **<u>ἐξῆλθον</u>** οἱ μαθηταὶ καὶ **<u>ἦλθον</u>** εἰς τὴν πόλιν καὶ εὗρον καθὼς **<u>εἶπεν</u>** αὐτοῖς, καὶ **<u>ἡτοίμασαν</u>** τὸ πάσχα.
And his disciples **<u>went out</u>**, and came into the city, and found things as he had **<u>said</u>** to them, and **<u>they prepared</u>** the Passover.

. . .

MARK 6:1

Καὶ **ἐξῆλθεν** ἐκεῖθεν, καὶ ἔρχεται εἰς τὴν πατρίδα αὐτοῦ, καὶ ἀκολουθοῦσιν αὐτῷ οἱ μαθηταὶ αὐτοῦ.

And **he went out** from there, and comes into his own country, and his disciples follow him.

REVELATION 16:20A

καὶ πᾶσα νῆσος **ἔφυγε**.

Every island **disappeared**.

MATTHEW 1:2

᾿Αβραὰμ **ἐγέννησε** τὸν ᾿Ισαάκ, ᾿Ισαὰκ δὲ **ἐγέννησε** τὸν ᾿Ιακώβ, ᾿Ιακὼβ δὲ **ἐγέννησε** τὸν ᾿Ιούδαν καὶ τοὺς ἀδελφοὺς αὐτοῦ.

Abraham **bore** Isaac, and Issac **bore** Jacob, and Jacob **bore** Judah and his brothers.

1 JOHN 2:26

Ταῦτα **ἔγραψα** ὑμῖν περὶ τῶν πλανώντων ὑμᾶς.

These things **I wrote** to you concerning those who would lead you astray.

LUKE 1:18A

Καὶ **εἶπεν** Ζαχαρίας πρὸς τὸν ἄγγελον, Κατὰ τί γνώσομαι τοῦτο;

And Zacharias **said** to the angel, "How shall

I know this?"

LUKE 1:63

καὶ αἰτήσας πινακίδιον **<u>ἔγραφεν</u>** λέγων, Ἰωάννης ἐστὶν ὄνομα αὐτοῦ.

He asked for a writing tablet, and **wrote**, "His name is John."

MARK 14:68B

καὶ **<u>ἐξῆλθεν</u>** ἔξω εἰς τὸ προαύλιον, καὶ ἀλέκτωρ **<u>ἐφώνησεν</u>**.

And **<u>he went out</u>** on the porch, and the rooster **<u>crowed</u>**.

136. **Recognition Drill**

Match the Greek with the translation. Answers are provided below.

a) They returned with the brother./b) I preached in their temple./c) You healed my brother./d) He proclaimed my good God./e) In the beginning was the Word./f) They shall be in their temple./g) We preached in our synagogue./h) Ye preached in the good temple./i) He healed your brethern./j) Then I loosed my servant.

_____ **1. τότε ἔλυσα τὸν δοῦλον μου.**

_____ **2. ἐθεράπευσεν τοὺς ἀδέλφους σου.**

_____ **3. ἐκρήυξατε ἐν τῷ ἀγαθῷ ἱερῷ.**

_____ **4. ἐκρήυξαμεν ἐν τῇ συναγωγῇ ἡμῶν.**

_____ **5. ἐκρήυξα ἐν τῷ ἱερῷ αὐτῶν.**

_____ **6. ὑπέστερψαν σὺν τῷ ἀδέλφῳ.**

_____ **7. ἐν ἀρχῇ ἦν ὁ Λόγος.**
_____ **8. ἔσονται ἐν τῷ ἱερῷ αὐτῶν.**
_____ **9. ἐκρήυξεν τὸν ἀγαθὸν θεόν μου.**
_____ **10. ἐθεράπευσας τὸν ἀδελφόν μου.**

Answers: 1j, 2i, 3h, 4g, 5b, 6a, 7e, 8f, 9d, 10c

TWENTY-FOUR

Λέγει αὐτῷ ὁ Ἰησοῦς· Ἐγώ εἰμι ἡ ὁδὸς καὶ ἡ ἀλήθεια καὶ ἡ ζωή· οὐδεὶς ἔρχεται πρὸς τὸν Πατέρα εἰ μὴ δι' ἐμοῦ.

Jesus says to him, "I am the way and the truth and the life. No one comes to the Father except through me."

John 14:6

Second Aorist Active and Middle Indicative

137. Vocabulary

γάρ for (conjunction) (Postpositive, i.e., it is the second element in a sentence or phrase.)

ἔβαλον I threw, I cast (second aorist of **βάλλω**)

ἐγενόμην I became (deponent second aorist of **γινώσκω**)

εἶδον I saw (second aorist of **βλέπω**; **εἶδον** is also considered to be the second aorist of **ὁράω** [I see]); in the GNT **εἶδον** normally has first aorist endings.

εἶπον I said (second aorist of **λέγω**)

ἔλαβον I took (second aorist of **λαμβάνω**)

ἤγαγον I led (second aorist of **ἄγω**)

ἦλθον I came, I went (second aorist of **ἔρχομαι**)

ἤνεγκα I bore, I brought (second aorist of **φέρω**); conjugated like the first aorist of **λύω**, but with **-κα** instead of **-σα**.)

λείπω I leave; the second aorist is **ἔλιπον**

ὄψομαι I shall see; (deponent future of **βλέπω**; **ὄψομαι** is also considered to be the future of **ὁράω**.)

πίπτω I fall; the second aorist is **ἔπεσον**

προσφέρω I bring to

WE HAVE ALREADY MENTIONED the fact that the second aorist is not a different tense from the first aorist, but a different way of forming the simple past (the preterit). Very few Greek verbs have both first and second aorist forms, just as very few verbs in English form the simple past by adding -ed and also by making changes within the body of the word. Consider the two English verbs, smell and drink. One forms its preterit by adding -ed (smelled) and the other undergoes a change within the body of the verb by means of a vowel change (drank). The two ways of forming this tense are different, but the result is the same in that both of these English verbs end up being in the simple past tense. Likewise in Greek. The uses of the second aorist are exactly the same as the uses of the first aorist. They are both simple past, but the second aorist usually has a vowel change in the verb form as seen in the vocabulary examples above.

. . .

UNFORTUNATELY, we cannot determine in advance whether any given Greek verb will have a first aorist or a second aorist form. In addition, we cannot determine beforehand what form that second aorist will take. The student must consult a lexicon to determine the forms.

THUS, as indicated above, the second aorist does not add **-σα** to the stem of the verb as the first aorist does. As already mentioned, the second aorist verb undergoes changes within the body of the verb. Examples:

The second aorist of **λαμβάνω** is **ἔλαβον**, **λαβ-** being the second aorist stem and **λαμβαν-** being the present stem. Note the augment.

βάλλω has a second aorist of **ἔβαλον**; **βαλ-** is the second aorist stem and **βαλλ-** is the present stem.

The second aorist stem is used to attach the endings for the second aorist active and middle tenses.

As in the case of the first aorist, the second aorist is a secondary tense. The second aorist, the first aorist, and the imperfect all add an augment to indicate past action, as the student has observed previously. The second aorist, being a secondary tense, adds secondary personal endings. Between these endings, just as in the case of the present and the imperfect, the variable vowel **ο** or **ε** is placed. The second aorist, therefore, is conjugated exactly like the imperfect, except that the imperfect is formed on the present stem, and the second aorist is formed on the second aorist stem.

138. Listed below are the forms of the second aorist active indicative of the verb **λείπω** (I leave). Notice that it has the **ε** at the beginning of each verb form to indicate past tense. It also

has a vowel change from **εί to ι**. Vowel changes are typical of second aorist verbs. In English, the verb "leave" also has a vowel change when it changes to "left" in the simple past.

SINGULAR

1P **ἔλιπον** I left
2P **ἔλιπες** you left
3P **ἔλιπε (ν)** he/she/it left

PLURAL

1P **ἐλίπομεν** we left
2P **ἐλίπετε** you (ye) left
3P **ἔλιπον** they left

NOTICE that the forms of the first person singular and the third person plural are identical. Analyzing the context of a verse will help the student determine whether **ἔλιπον** means "I left" or "they left."

139. **Recognition Drill**

Match the Greek with the translation. Answers are provided below.

a) He proclaimed your good God./b) They left me in the good temple./c) Ye left with the brother./d) We will see him in your temple./e) You healed my brother./f) I came with their brother.g) She shall see you in their temple./h) I shall see your brothers/.i) You left him in the synagogue./j) I saw your servant.

_____ **1. εἶδον τὸν δοῦλον σου.**

_____ 2. ὄψομαι τοὺς ἀδέλφους σου.
_____ 3. ἔλιπον με ἐν τῷ ἀγαθῷ ἱερῷ.
_____ 4. ἔλιπες ἀυτὸν ἐν τῇ συναγωγῇ.
_____ 5. ὄψεται σε ἐν τῷ ἱερῷ αὐτῶν.
_____ 6. ἐλίπετε σὺν τῷ ἀδέλφῳ.
_____ 7. ἦλθον σὺν τῷ ἀδέλφῳ αὐτῶν.
_____ 8. ὄψομεθα ἀυτὸν ἐν τῷ ἱερῷ σου.
_____ 9. ἐκρήυξεν τὸν ἀγαθὸν θεόν σου.
_____ 10. ἐθεράπευσας τὸν ἀδελφόν μου.

Answers: 1j, 2h, 3b, 4i, 5g, 6c, 7f, 8d, 9a, 10e

TWENTY-FIVE

πιστεύετέ μοι ὅτι ἐγὼ ἐν τῷ Πατρὶ καὶ ὁ Πατὴρ ἐν ἐμοί.
Believe me that I am in the Father and the Father is in Me.
John 14:11a

Review: Additional Greek Verses Containing the Aorist

140. Aorist verbs, as well as the English equivalents, are bolded and underlined.

LUKE 1:13A

<u>εἶπεν</u> δὲ πρὸς αὐτὸν ὁ ἄγγελος, Μὴ φοβοῦ.
And the angel **<u>said</u>** to him, "Do not fear."

LUKE 2:30

ὅτι **<u>εἶδον</u>** οἱ ὀφθαλμοί μου τὸ σωτήριόν σου.
Because my eyes **<u>saw</u>** your salvation.

. . .

LUKE 1:30

καὶ **εἶπεν** ὁ ἄγγελος αὐτῇ, Μὴ φοβοῦ, Μαριάμ, **εὗρες** γὰρ χάριν παρὰ τῷ θεῷ.·

And the angel **said,** "Do not fear, Mary, you have **found** favor with God."

LUKE 1:34

εἶπεν δὲ Μαριὰμ πρὸς τὸν ἄγγελον, Πῶς ἔσται τοῦτο, ἐπεὶ ἄνδρα οὐ γινώσκω;

Mary **said** to the angel, "How can this be, seeing I am a virgin?"

LUKE 1:38

εἶπεν δὲ Μαριάμ, Ἰδοὺ ἡ δούλη κυρίου· γένοιτό μοι κατὰ τὸ ῥῆμά σου. καὶ **ἀπῆλθεν** ἀπ' αὐτῆς ὁ ἄγγελος.

Mary **said**, "Behold, the servant of the Lord; let it be done to me according to your word." And the angel **left** her.

LUKE 2:1

Ἐγένετο δὲ ἐν ταῖς ἡμέραις ἐκείναις **ἐξῆλθεν** δόγμα παρὰ Καίσαρος Αὐγούστου.

Now it **happened** that in those days, a decree **went out** from Caesar Augustus.

LUKE 2:27

καὶ **ἦλθεν** ἐν τῷ πνεύματι εἰς τὸ ἱερόν.

He **came** in the Spirit into the temple.

. . .

LUKE 2:30

ὅτι **εἶδον** οἱ ὀφθαλμοί μου τὸ σωτήριόν σου.

For my eyes **saw** your salvation.

LUKE 2:41

Καὶ **ἐπορεύοντο** οἱ γονεῖς αὐτοῦ κατ' ἔτος εἰς Ἰερουσαλὴμ τῇ ἑορτῇ τοῦ πάσχα.

His parents **went** every year to Jerusalem at the feast of the Passover.

MARK 1:28

καὶ **ἐξῆλθεν** ἡ ἀκοὴ αὐτοῦ εὐθὺς εἰς ὅλην τὴν περίχωρον τῆς Γαλιλαίας.

And his fame **spread** immediately through all the region of Galilee.

MARK 2:13 (ἐδίδασκεν is imperfect)

Καὶ **ἐξῆλθε** πάλιν παρὰ τὴν θάλασσαν· καὶ πᾶς ὁ ὄχλος ἤρχετο πρὸς αὐτόν, καὶ ἐδίδασκεν αὐτούς.

And then **he went out** again by the sea; and all the multitude came to him and he was teaching them.

MARK 3:1

Καὶ **εἰσῆλθε** πάλιν εἰς τὴν συναγωγήν· καὶ ἦν ἐκεῖ ἄνθρωπος ἐξηραμμένην ἔχων τὴν

χεῖρα.
And **he entered** into the synagogue again, and a man was there who had a withered hand.

JOHN 3:22 (διέτριβε and ἐβάπτιζεν are imperfect)
Μετὰ ταῦτα **ἦλθεν** ὁ Ἰησοῦς καὶ οἱ μαθηταὶ αὐτοῦ εἰς τὴν Ἰουδαίαν γῆν, καὶ ἐκεῖ διέτριβε μετ᾽ αὐτῶν καὶ ἐβάπτιζεν.
After these things Jesus and his disciples **came** into the land of Judea, and there he remained with them and was baptizing.

1 JOHN 2:14 (ἐγνώκατε is the perfect tense)
ἔγραψα ὑμῖν, παιδία, ὅτι ἐγνώκατε τὸν πατέρα. **ἔγραψα** ὑμῖν, πατέρες, ὅτι ἐγνώκατε τὸν ἀπ᾽ ἀρχῆς. **ἔγραψα** ὑμῖν, νεανίσκοι, ὅτι ἰσχυροί ἐστε καὶ ὁ λόγος τοῦ Θεοῦ ἐν ὑμῖν μένει καὶ νενικήκατε τὸν πονηρόν.
I **wrote** to you, children, because you have known the father. I **wrote** to you, fathers, because you have known Him (who) is from the beginning. I **wrote** to you, young men, because you are strong and the word of God abides in you and you have overcome the wicked one.

TWENTY-SIX

Ἐὰν ἀγαπᾶτέ με, τὰς ἐντολὰς τὰς ἐμὰς τηρήσετε.
If you love Me, keep My commandments.
John 14:15

Aorist Passive Indicative. Future Passive Indicative

141. Vocabulary

ἀναλαμβάνω I take up

ἐβλήθην I was thrown, I was cast; aorist passive of **βάλλω**

ἐγενήθην I became; aorist passive of **γίνομαι**

ἐγνώσθην I was known; aorist passive of **γινώσκω**

ἐδιδάχθην I was taught; aorist passive of **διδάσκω**

ἐκηρήχθην I was preached, I was proclaimed; aorist passive of **κηρύσσω**

ἐλήμφθην I was taken; aorist passive of **λαμβάνω**

ἐπορεύθην I went; aorist passive of **πορεύομαι**

ἠγέρθην I was risen up; aorist passive of **ἐγείρω** (rise, raise up)

ἠκούσθην I was heard; aorist passive of **ἀκούω**

ἠνέχθην I was brought; aorist passive of **φέρω**

ἤχθην I was led; aorist passive of **ἄγω**

ὤφθην I was seen; aorist passive of **βλέπω** or **ὁράω**

IN LEARNING to recognize these forms, notice the similarities between each infinitive and its aorist passive form.

THE AORIST PASSIVE indicative and the future passive indicative are formed from the aorist passive stem, which is formed by adding **θε** to the stem of the verb. Throughout the indicative, the **θε** is lengthened to **θη**. Therefore, the aorist passive stem of **λύω** is **λυθη**.

THE AORIST PASSIVE is a secondary tense, so the augment is added before the tense stem is added and before the secondary personal endings are added. The active personal endings are added. Note that the endings in the active of the secondary tenses are: .**-ν**, **-ς**, none, **-μεν**, **-τε**, **-ν** (or **-σαν**). These are the same endings that are employed in the imperfect active indicative except that in the third person plural the ending is **-σαν** and not **-ν.** No variable vowel is employed; the endings are added directly to the tense stem **θη**.

IN THE FUTURE passive indicative **θη** is added to the verb stem, followed by **-σο** or **-σε**, and then by the appropriate middle primary ending. The future is a primary and not a secondary tense, so no augment is used in the future passive indicative. Thus, the future passive indicative is formed exactly like the future middle except that the stem is **λυθησο/ε**, instead of **λυσο/ε**.

142. The aorist passive indicative of **λύω** is as follows. Note the secondary active endings: **-ν**, **-ς**, none, **-μεν**, **-τε**, **-σαν**.

SINGULAR

1P **ἐλύθην** I was loosed
2P **ἐλύθης** you were loosed
3P **ἐλύθη** he/she/it was loosed

PLURAL

1P **ἐλύθημεν** we were loosed
2P **ἐλύθητε** you (ye) were loosed
3P **ἐλύθησαν** they were loosed

REMEMBER that with deponent verbs which have aorist passive forms, the meaning will be aorist active in nature, so that the aorist passive of **γίνομαι** (I become) is **ἐγενήθην** (I became).

Another example can be seen with the verb **πορεύομαι** (I go), whose aorist passive form, **ἐπορεύθην**, means "I went."

143. The future passive indicative of **λύω** is given below. Note that, as expected, the endings are middle/passive in nature.

. . .

SINGULAR

1P **λυθήσομαι** I will be loosed
2P **λυθήσῃ** you will be loosed
3P **λυθήσεται** he/she/it will be loosed

PLURAL

1P **λυθησόμεθα** we will be loosed
2P **λυθήσεσθε** you (ye) will loosed
3P **λυθήσονται** they will be loosed

144. Thus far we have studied a number of verb forms and tenses. The following chart, featuring the verb **λύω**, may be useful as a summary of the indicative forms.

Present Active **λύω** "I loose."
Present Middle **λύομαι** "I loose for myself."
Present Passive **λύομαι** "I am being loosed."
Imperfect Active **ἔλυον** "I was loosing."
Imperfect Middle **ἐλυόμην** "I was loosing for myself."
Imperfect Passive **ἐλυόμην** "I was being loosed."
Future Active **λύσω** "I shall (will) loose."
Future Middle **λύσομαι** "I will loose for myself."
Future Passive **λυθήσομαι** "I will be loosed."
Aorist Active **ἔλυσα** "I loosed."
Aorist Middle **ἐλυσάμην** "I loosed for myself."
Aorist Passive **ἐλύθην** "I was loosed."

FORMATION OF AORIST **Passive Stems Whose Verb Stems End in a Consonant**

. . .

145. Before the **θ** of the aorist passive tense-suffix (**θη**), a final **π** or **β** of the verb stem is changed to **φ**, a final **κ** or **γ** is changed to **χ**, and a final **τ**, **δ**, or **θ** is changed to **σ**. The changes in the case of **π**, **β**, **κ**, or **γ** can be remembered if the student observes that **θ** is equivalent to "th" and that the changes entail adding an "h" to the preceding letters to make them conform to the "th" pattern. Thus, before "th," "p," becomes "ph" and "k" or "g" become "chi."

Examples: The aorist passive of **πέμπω** is **ἐπέμφθην**, of **ἄγω**, **ἤχθην** and of **πείθω**, **ἐπείσθην**.

VERSES WITH THE AORIST PASSIVE. Applicable verb examples are bolded and underlined.

MARK 2:1

Καὶ εἰσελθὼν πάλιν εἰς Καφαρναοὺμ δι' ἡμερῶν **<u>ἠκούσθη</u>** ὅτι ἐν οἴκῳ ἐστίν.

And (as he was) entering into Capernaum a few days later, **<u>it was rumored</u>** that he was (is) in a house.

MATTHEW 1:16

Ἰακὼβ δὲ ἐγέννησεν τὸν Ἰωσὴφ τὸν ἄνδρα Μαρίας, ἐξ ἧς **<u>ἐγεννήθη</u>** Ἰησοῦς ὁ λεγόμενος Χριστός.

Jacob became the father of Joseph, the husband of Mary, from whom **<u>was born</u>** Jesus, who is called Christ.

. . .

146. **Recognition Drill**

Match the Greek with the translation. Answers are provided below.

a) I was heard by them./b) The book was written by them./c) I was taught by him./d) We were carried by you./e) I will be loosed by the man./f) Ye were seen by the men./g) I was carried by them./h) The book was carried by them./i) I was taught by him./j) They were written by you.

_____ **1. ἐγράφησαν ὑπὸ σου.**

_____ **2. ἠνέχθημεν ὑπὸ σου.**

_____ **3. ἐδιδάχθην ὑπ' αὐτοῦ.**

_____ **4. ἠνέχθην ὑπ' αὐτῶν.**

_____ **5. τὸ βιβλίον ἐγράφη ὑπ'αὐτῶν.**

_____ **6. ἐδιδάχθην ὑπ' αὐτοῦ.**

_____ **7. ἠκούσθην ὑπ' αὐτῶν.**

_____ **8. λυθήσομαι ὑπὸ τοῦ ἀνθρώπου.**

_____ **9. ὤφθητε ὑπὸ τῶν ἀνθρώπων.**

_____ **10. τὸ βιβλίον ἤχθη ὑπ'αὐτῶν.**

Answers: 1j, 2d, 3i, 4g, 5b, 6c, 7a, 8e, 9f, 10h

TWENTY-SEVEN

ἤδη ὑμεῖς καθαροί ἐστε διὰ τὸν λόγον ὃν λελάληκα ὑμῖν.
You are already pure because of the Word which I have spoken to you.
John 15:3

Present Participles

147. Vocabulary

προσέρχομαι deponent verb, I come, I go to (takes the dative case)

148. The participle is very important in understanding Greek. It is found in all tenses except the imperfect ("I was loosing") and the pluperfect ("I had loosed"). It occurs in all 3 voices

(active, middle and passive). Here is the declension of the present active participle of **λύω** (loosing):

SINGULAR

N. (M) **λύων** (F) **λύουσα** (N) **λῦον**
G. (M) **λύοντος** (F) **λυούσης** (N) **λύοντος**
D. (M) **λύοντι** (F) **λυούσῃ** (N) **λύοντι**
A. (M) **λύοντα** (F) **λύουσαν** (N) **λύον**

PLURAL

N. (M) **λύοντες** (F) **λύουσαι** (N) **λύοντα**
G. (M) **λυόντων** (F) **λυουσῶν** (N) **λυόντων**
D. (M) **λύουσι(ν)** (F) **λυούσαις**
(N) **λύουσι(ν)**
A. (M) **λύοντας** (F) **λυούσας** (N) **λύοντας**

PLEASE KEEP in mind that these present active participles, as well as the present middle participles listed below, will usually be used in combination with the definite article and a noun, all of which will usually have the same endings. If the student has mastered the definite article and knows the meaning of each, he or she will have little trouble with the endings of the participles listed above.

NOTE that the masculine and neuter participle forms are formed according to the third declension. The masculine is exactly like the noun **ἄρχων**. The feminine is declined just like the first declension noun **δόξα**.

. . .

149. Listed below is the declension of the present middle participle **λυόμενος**, **η**, **ον**, (loosing for himself), and the present passive participle **λυόμενος**, **η**, **ον**, (being loosed). The latter is applicable to most occurrences in the Greek New Testament. Notice that the forms for both the middle and passive participle systems are exactly the same even though the meanings are quite different.

SINGULAR

N. (M) **λυόμενος** (F) **λυομένη**
(N) **λυόμενον**
G. (M) **λυομένου** (F) **λυομένης**
(N) **λυομένου**
D. (M) **λυομένῳ** (F) **λυομένῃ**
(N) **λυομένῳ**
A. (M) **λυόμενον** (F) **λυομένην**
(N) **λυόμενον**

PLURAL

N. (M) **λυόμενοι** (F) **λυόμεναι**
(N) **λυόμενα**
G. (M) **λυομένων** (F) **λυομένων**
(N) **λυομένων**
D. (M) **λυομένοις** (F) **λυομέναις**
(N) **λυομένοις**
A. (M) **λυομένους** (F) **λυομένας**
(N) **λυόμενα**

THIS DECLENSION IS like that of adjectives of the second

and first declension. The present participles are formed on the present stem of the verb.

150. **The Use of Participles**

Participles are verbal adjectives, and as such, have gender, number and case. Like other adjectives they agree in gender, number and case with the nouns they modify. Since a participle partakes of the nature of verbs, it has tense and voice and can take a direct object if it is the participle of a transitive verb.

151. **The Present Participle of εἰμί (being)**

THE DECLENSION OF "BEING," the present participle of **εἰμί** (I am), is as follows:

SINGULAR

N. (M) **ὤν** (F) **οὖσα** (N) **ὄν**
G. (M) **ὄντος** (F) **οὔσης** (N) **ὄντος**
D. (M) **ὄντι** (F) **οὔσῃ** (N) **ὄντι**
A. (M) **ὄντα** (F) **οὖσαν** (N) **ὄν**

PLURAL

N. (M) **ὄντες** (F) **οὖσαι** (N) **ὄντα**
G. (M) **ὄντων** (F) **οὐσῶν** (N) **ὄντων**
D. (M) **οὖσι(ν)** (F) **οὔσαις** (N) **οὖσι(ν)**
A. (M) **ὄντας** (F) **οὔσας** (N) **ὄντας**

. . .

THIS DECLENSION HAS the same endings as the present active participle of **λύω**, listed above in section 149.

152. **Greek Verses Containing Participles**

Participles as well as the English equivalents, are bolded and underlined.

MARK 8:18A

ὀφθαλμοὺς **<u>ἔχοντες</u>** οὐ βλέπετε καὶ ὦτα **<u>ἔχοντες</u>** οὐκ ἀκούετε;

You **<u>have</u>** eyes and (yet) you do not see and you **<u>have</u>** ears and you do not hear?

MARK 1: 22

καὶ ἐξεπλήσσοντο ἐπὶ τῇ διδαχῇ αὐτοῦ, ἦν γὰρ **<u>διδάσκων</u>** αὐτοὺς ὡς ἐξουσίαν **<u>ἔχων</u>**, καὶ οὐχ ὡς οἱ γραμματεῖς.

And they were amazed at his teaching, for he was **<u>teaching</u>** them as one **<u>having</u>** authority and not as the scribes.

MARK 2:3

καὶ ἔρχονται πρὸς αὐτὸν παραλυτικὸν **<u>φέροντες</u>**, **<u>αἰρόμενον</u>** ὑπὸ τεσσάρων.

And they come **<u>bringing</u>** to him a paralytic, **<u>being carried</u>** by four (people).

. . .

MARK 2:6

ἦσαν δέ τινες τῶν γραμματέων ἐκεῖ **καθήμενοι** καὶ **διαλογιζόμενοι** ἐν ταῖς καρδίαις αὐτῶν.·

And some of the scribes were **sitting** there and **reasoning** in their hearts.

JOHN 1:29

Τῇ ἐπαύριον βλέπει ὁ Ἰωάννης τὸν Ἰησοῦν **ἐρχόμενον** πρὸς αὐτὸν καὶ λέγει, ἴδε ὁ ἀμνὸς τοῦ Θεοῦ ὁ **αἴρων** τὴν ἁμαρτίαν τοῦ κόσμου.

The next day he sees Jesus **coming** toward him and he says, "Behold the Lamb of God who takes away (the one **taking away**) the sin of the world."

REVELATION 18:1

Μετὰ ταῦτα εἶδον ἄλλον ἄγγελον **καταβαίνοντα** ἐκ τοῦ οὐρανοῦ, **ἔχοντα** ἐξουσίαν μεγάλην, καὶ ἡ γῆ ἐφωτίσθη ἐκ τῆς δόξης αὐτοῦ.

After these things I saw another angel **coming down** out of heaven, **having** great authority, and the earth was illuminated with his glory.

1 JOHN 2:10

ὁ **ἀγαπῶν** τὸν ἀδελφὸν αὐτοῦ ἐν τῷ φωτὶ

μένει, καὶ σκάνδαλον ἐν αὐτῷ οὐκ ἔστιν. The person **loving** his brother abides in the light, and there is no cause for stumbling in him.

TWENTY-EIGHT

μείνατε ἐν ἐμοί, κἀγὼ ἐν ὑμῖν.
Remain in Me and I in you.
John 15:4a

Intensive Recognition Practice with Literal English Translation

153. English words not in the Greek text, but needed for clarity, are supplied in parentheses. The word for word translation may seem stilted in some cases, but the student will be able to follow along one word at a time. By now the student will notice that he or she recognizes most of the Greek words used.

JOHN 1: 1 – 2

Ἐν ἀρχῇ ἦν ὁ Λόγος, καὶ ὁ Λόγος ἦν πρὸς

τὸν Θεόν, καὶ Θεὸς ἦν ὁ Λόγος. Οὗτος ἦν ἐν ἀρχῇ πρὸς τὸν Θεόν.

In (the) beginning was the Word, and the Word was with the God, and God was the Word.

JOHN 15:1

Ἐγώ εἰμι ἡ ἄμπελος ἡ ἀληθινή, καὶ ὁ πατήρ μου ὁ γεωργός ἐστι.

I am the vine, the true (one), and the father of me (my father) the husbandman is.

JOHN 14:6

λέγει αὐτῷ ὁ Ἰησοῦς, ἐγώ εἰμι ἡ ὁδὸς καὶ ἡ ἀλήθεια καὶ ἡ ζωή. οὐδεὶς ἔρχεται πρὸς τὸν πατέρα εἰ μὴ δι᾽ ἐμοῦ.

Says to him the Jesus, "I am the way and the truth and the life. No one comes to the Father except through me."

JOHN 6:48

ἐγώ εἰμι ὁ ἄρτος τῆς ζωῆς.

I am the bread of the life.

JOHN 10:30

ἐγὼ καὶ ὁ πατὴρ ἕν ἐσμεν.

I and the Father one are.

. . .

JOHN 5:41

καὶ οὐ θέλετε ἐλθεῖν πρός με ἵνα ζωὴν ἔχητε.

And not ye want to come to me, in order that life ye might have.

MARK 4:14

ὁ σπείρων τὸν λόγον σπείρει.

The one sowing the word, sows.

REVELATION 7:12

... λέγοντες· ἀμήν, ἡ εὐλογία καὶ ἡ δόξα καὶ ἡ σοφία καὶ ἡ εὐχαριστίακαὶ ἡ τιμὴ καὶ ἡ δύναμις καὶ ἡ ἰσχὺς τῷ Θεῷ ἡμῶν εἰς τοὺς αἰῶνας τῶν αἰώνων. ἀμήν.

... saying, "Amen, the blessing and the glory and the wisdom and the thanks (and) honor and power and strength (be) to the God of us unto the ages of the ages, Amen."

JOHN 8:15

ὑμεῖς κατὰ τὴν σάρκα κρίνετε, ἐγὼ οὐ κρίνω οὐδένα.

You judge according to the flesh. I judge no one.

1 JOHN 5:12

ὁ ἔχων τὸν υἱὸν ἔχει τὴν ζωήν. ὁ μὴ ἔχων τὸν υἱὸν τοῦ θεοῦ τὴν ζωὴν οὐκ ἔχει.

He who has the Son has the life. The one having not the Son of the God the life not has.

JOHN 1:29

Τῇ ἐπαύριον βλέπει ὁ Ἰωάννης τὸν Ἰησοῦν ἐρχόμενον πρὸς αὐτὸν καὶ λέγει, ἴδε ὁ ἀμνὸς τοῦ Θεοῦ ὁ αἴρων τὴν ἁμαρτίαν τοῦ κόσμου.

The next day sees the John the Jesus coming toward him and says, "Behold the Lamb of the God, the One taking away the sin of the world.

JOHN 2:3

καὶ ὑστερήσαντος οἴνου λέγει ἡ μήτηρ τοῦ Ἰησοῦ πρὸς αὐτόν, οἶνον οὐκ ἔχουσι.

And being used up (the) wine, says the mother of the Jesus to Him, "Wine not they have."

REVELATION 1:8

Ἐγώ εἰμι τὸ Α καὶ τὸ Ω, λέγει Κύριος ὁ Θεός, ὁ ὢν καὶ ὁ ἦν καὶ ὁ ἐρχόμενος, ὁ παντοκράτωρ.

"I am the Alpha and the Omega," says (the) Lord the God, the One being and the One (Who) was and the One coming, the Almighty.

. . .

JOHN 6:8

λέγει αὐτῷ εἷς ἐκ τῶν μαθητῶν αὐτοῦ,
Ἀνδρέας ὁ ἀδελφὸς Σίμωνος Πέτρου.
Says to Him, one of the disciples of Him,
Andrew, the brother of Simon Peter.

JOHN 6:20

ὁ δὲ λέγει αὐτοῖς· ἐγώ εἰμι, μὴ φοβεῖσθε.
And He says to them. "I am, not be afraid."

JOHN 6:33

ὁ γὰρ ἄρτος τοῦ Θεοῦ ἐστιν ὁ καταβαίνων
ἐκ τοῦ οὐρανοῦ καὶ ζωὴν διδοὺς τῷ κόσμῳ.
For the bread of the God is the One coming
down out of the heaven and life having
given to the world.

JOHN 6:34

εἶπον οὖν πρὸς αὐτόν, Κύριε, πάντοτε δὸς
ἡμῖν τὸν ἄρτον τοῦτον.
They said therefore to him, "Lord, always
give us the bread this.

JOHN 6:35

εἶπε δὲ αὐτοῖς ὁ Ἰησοῦς, ἐγώ εἰμι ὁ ἄρτος
τῆς ζωῆς. ὁ ἐρχόμενος πρός με οὐ μὴ
πεινάσῃ, καὶ ὁ πιστεύων εἰς ἐμὲ οὐ μὴ
διψήσῃ πώποτε.
Said to them the Jesus, "I am the bread of the

life. The one coming to me by no means hungers, and the one believing in me by no means thirsts ever.

JOHN 6:42

καὶ ἔλεγον, οὐχ οὗτός ἐστιν Ἰησοῦς ὁ υἱὸς Ἰωσήφ, οὗ ἡμεῖς οἴδαμεν τὸν πατέρα καὶ τὴν μητέρα; πῶς οὖν λέγει οὗτος ὅτι ἐκ τοῦ οὐρανοῦ καταβέβηκα;

And they were saying, "Not this (man) is Jesus, the son of Joseph, not we know the father and the mother? How therefore says this (man) that out of heaven he has come down?"

JOHN 6:47-8

ἀμὴν ἀμὴν λέγω ὑμῖν, ὁ πιστεύων εἰς ἐμὲ ἔχει ζωὴν αἰώνιον. ἐγώ εἰμι ὁ ἄρτος τῆς ζωῆς.

Truly, truly I say to you. The one believing in me has life eternal. I am the bread of the life.

JOHN 6:51

ἐγώ εἰμι ὁ ἄρτος ὁ ζῶν ὁ ἐκ τοῦ οὐρανοῦ καταβάς. ἐάν τις φάγῃ ἐκ τούτου τοῦ ἄρτου, ζήσεται εἰς τὸν αἰῶνα. καὶ ὁ ἄρτος δὲ ὃν ἐγὼ δώσω, ἡ σάρξ μού ἐστιν, ἣν ἐγὼ δώσω ὑπὲρ τῆς τοῦ κόσμου ζωῆς.

I am the bread, the living, the (bread) out of heaven having come down. If anyone eats out of this, the bread, he shall live forever. And the bread which I shall give, the flesh of me is. Which I shall give for the life of the world.

JOHN 8:13

εἶπον οὖν αὐτῷ οἱ Φαρισαῖοι, σὺ περὶ σεαυτοῦ μαρτυρεῖς. ἡ μαρτυρία σου οὐκ ἔστιν ἀληθής.

Said therefore to Him the Pharisees, "You concerning yourself testify. The testimony of you not is true."

JOHN 8:23

καὶ εἶπεν αὐτοῖς, ὑμεῖς ἐκ τῶν κάτω ἐστέ, ἐγὼ ἐκ τῶν ἄνω εἰμί. ὑμεῖς ἐκ τοῦ κόσμου τούτου ἐστέ, ἐγὼ οὐκ εἰμὶ ἐκ τοῦ κόσμου τούτου.

And He said to them, "You out of below are, I out of above am. You out of the world this are, I not am out of the world this."

JOHN 8:34

ἀπεκρίθη αὐτοῖς ὁ Ἰησοῦς, ἀμὴν ἀμὴν λέγω ὑμῖν ὅτι πᾶς ὁ ποιῶν τὴν ἁμαρτίαν δοῦλός ἐστι τῆς ἁμαρτίας.

Answered them the Jesus, "Truly, truly I say to you that everyone doing the sin slave is of sin."

. . .

JOHN 8:45

ἐγὼ δὲ ὅτι τὴν ἀλήθειαν λέγω, οὐ πιστεύετέ μοι.

(And) I, because the truth I speak, not you believe me.

JOHN 12:12

Τῇ ἐπαύριον ὄχλος πολὺς ὁ ἐλθὼν εἰς τὴν ἑορτήν, ἀκούσαντες ὅτι ἔρχεται Ἰησοῦς εἰς Ἱεροσόλυμα ...

The next day a crowd large, the (one) coming to the feast, having heard that is coming Jesus into Jerusalem ...

TWENTY-NINE

Εἰρήνην ἀφίημι ὑμῖν, εἰρήνην τὴν ἐμὴν δίδωμι ὑμῖν.
Peace I leave with you, My own peace I give to you.
John 14:27a

A Reference Guide to Basic Verb Concepts: Keys to Recognition

154. Verb Tenses

This chapter and chapter 30 are intended to be reference guides in which the grammatical concepts covered in this book are summarized. The concepts may seem complicated at first, but as the student gets more familiar with the Greek text and its grammar, the summaries presented will become more clear. If the student has been diligently studying the material presented thus far in the book, the concepts below will present little difficulty.

There are seven tenses in Greek: the present (I loose), imperfect (I was loosing), future (I will loose), aorist (I loosed), perfect (I have loosed), pluperfect (I had loosed), and future perfect (I will/shall have loosed). Let's look at these tenses one by one.

The present tense describes an ongoing or continuous action at the present time. For example, **λύω**, which means "I loose," or "I am loosing," or "I do loose." Examples in English would be, "I eat my lunch," or "I am eating my lunch," or "I do eat my lunch."

The imperfect tense describes an ongoing or continuous action in the past. For example, **ἔλυον** means "I was loosing." An example in English would be, "When I got home, she was eating her lunch."

The future describes a simple or ongoing action in the future. For example, **λύσω** means, "I shall (will) loose." English examples would be, "When I get home, I will eat my lunch," or "From now on I will visit Paris every single year."

The aorist tense is used to describe a simple action in the past, that is, a completed action occurring at one point in time in the past. For example, **ἔλυσα** means "I loosed," An English example: "I ate my lunch at the restaurant."

The perfect tense is used to describe a completed action in the present time. For example, **λέλυκα**, which means, "I have loosed." An English example would be: "I have stopped eating lunch at that restaurant." Because this Greek tense is rare, it is not discussed in depth in this book. It is easily recognizable, however, because it features a "reduplication" of the first consonant of the verb. With the verb **λύω**, this reduplication would be **λὲλ**, featuring an **ε** between the two occurrences of **λ**. This tense also features the same endings as the aorist, except **κ** stands in place of the **σ** of the aorist.

The pluperfect describes a completed action in the past,

and like the perfect tense, is rather rare and is not discussed in depth in this book. An example of this tense is **λέλυκειν**, which means "I had loosed." An English example would be: "By the time she got home, I had gone to lunch." Like the perfect tense, the pluperfect has the reduplication. It has been described as the perfect tense adjusted backward in time.

The future perfect is used to describe a completed action in the future. This tense is also not discussed in this book due to its rarity. "I shall have loosed," would be the English equivalent.

155. **Primary and Secondary Tenses**

The present, future, and perfect are called primary tenses. As mentioned above, the perfect tense is not discussed in this book.

Included in this category in the present tense are the present active indicative, the present middle indicative and the present passive indicative.

Included in this category in the future tense are the future active indicative, the future middle indicative and the future passive indicative.

Included in this category in the perfect tense are the perfect active indicative and the perfect passive indicative.

The imperfect, aorist and pluperfect are called secondary tenses.

Included in this category in the imperfect tense are the imperfect active, middle and passive indicative.

Included in this category in the aorist tense are the aorist active, middle and passive indicative.

Included in this category in the pluperfect tense is the pluperfect active indicative. As mentioned above, the pluperfect is not discussed in this book.

All of these tenses are grouped together in their respective categories because they have certain features in common regarding the personal verb endings they receive.

156. Study the following chart, which has been repeated from chapter twenty-six. Remember, the present, future, and perfect are the primary tenses and the imperfect and aorist are the secondary tenses.

Present Active **λύω** "I loose."
Present Middle **λύομαι** "I loose for myself."
Present Passive **λύομαι** "I am being loosed."
Imperfect Active **ἔλυον** "I was loosing."
Imperfect Middle **ἐλυόμην** "I was loosing for myself."
Imperfect Passive **ἐλυόμην** "I was being loosed."
Future Active **λύσω** "I shall (will) loose."
Future Middle **λύσομαι** "I will loose for myself."
Future Passive **λυθήσομαι** "I will be loosed."
Aorist Active **ἔλυσα** "I loosed."
Aorist Middle **ἐλυσάμην** "I loosed for myself."
Aorist Passive **ἐλύθην** "I was loosed."

IN THE PRIMARY TENSES, the personal verb endings in the active are as follows: **-ω**, -**εις**, -**ει**, -**ομεν**, -**ετε**, -**ουσι(ν)**. For example, see the present active indicative and the future active indicative in the chart above. Only the first-person singular of the verb forms is listed.

In the primary tenses, the personal verb endings in the middle and passive are: **-μαι**, -**σαι** (shortened to **ῃ**), -**ται**, -**μεθα**, -**σθε**, -**νται**. For examples, see the present middle indicative, the future middle indicative, and the future passive indicative in the chart, above.

. . .

IN THE SECONDARY TENSES, the personal verb endings in the active are -**ν**, **ς**, none, -**μεν**, -**τε**, -**ν** (or -**σαν**). For examples, see the present middle indicative, the future middle indicative, the imperfect active indicative, the aorist active indicative, and the aorist passive indicative in the chart, above.

In the secondary tenses, the personal verb endings in the imperfect middle indicative, the imperfect passive indicative and the aorist middle indicative are as follows: -**μην**, -**σο** (shortened to **ου**), -**το**, -**μεθα**, -**σθε**, -**ντο**. Refer to the chart, above.

THIRTY

Summary of Recognition Tips: A Reference Guide

157. By this time the student should be able to understand the many portions of Scripture cited at the top of each chapter. Take another look and see how you do.

WE HAVE DISCUSSED many ways to recognize words, phrases and sentences in the GNT. They are summarized below.

1. Utilize your knowledge of cognates, in other words terms that are similar or identical in both English and Greek. For example, the Greek word **ἀπόστολος** reminds us of the similar English word "apostle."

2. Recognizing and knowing the forms of the definite articles is essential to comprehending the meaning of the Greek text, because the adjectives and nouns used with them will almost always have identical endings.

EXAMPLE ONE: consider the following sentence, which features three words that are in the accusative case: a feminine singular definite article, an adjective and a feminine singular noun. **ὁ μαθητὴς γινώσκει τὴν ἕτεραν ἐκκλησίαν.** ("The disciple knows the other church.") Each **ν** has been underlined above. Do you recognize the consistencies of the endings on the definite article, the adjective, and the noun? More often than not, the definite article, adjective and nouns will have the same endings. In other words, the endings of the definite articles, adjectives and nouns agree in gender, number and grammatical case, which is indicated by those endings. Even if the endings on the nouns and/or adjectives do not agree with the definite article, the form of that definite article will guide you into understanding the meaning of the sentence.

EXAMPLE TWO: **ὁ μαθητὴς βλέπει τὰς ἐσχατὰς ὥρας.** ("The disciple sees the last hours.") This sentence features a feminine plural definite article, a feminine plural adjective and a feminine plural noun, all of which are direct objects. The consistencies have been underlined.

EXAMPLE THREE: **ἕχομεν τὸ δῶρον τῶν πονηρῶν ἀδέλφων.** ("We have the gift of the evil brethren.") Here we have the genitive case affecting a masculine plural definite article, adjective and noun. Again, the consistencies have been underlined.

CONSULT the three noun declensions covered in this book for similar consistencies.

Of special note is the dative singular of all declensions, whose nouns end in **ι**. Two examples are **συναγωγῇ** and **πνεύματι**. The sign of the dative singular, then, is this final **ι**. With **συναγωγῇ** the final **ι** is under **η** as **ῇ**. This is called the **ι**-subscript.

3. Pay attention to the endings on verbs, participles and nouns, as they give clues to meaning.

a. The verbs ending in **-ω** as in **ἀκούω** and **μαι** as in **ἀποκρίνομαι** and **-μην** as in **ἐλυόμην**, will signal the first-person singular, or "I," form.
b. The verbs ending in **-εις** as in **ἀκούεις** will indicate the second person singular or "thou" form.
c. The verbs ending in **-ει** as **ἀκούει** and **-ται**, as in **ἀποκρίνεται** and **-το,** as in **ἐλύετο**, will signal the third person singular or the "he, she, it" forms.
d. The verb endings **-μεν** as in **ἀκούομεν** and **-μεθα** as in **ἀποκρινόμεθα** will indicate that the verb is first person plural and would be translated as "we."
e. The verb endings **-τε** as in **ἀκούετε** and **-σθε** as in **ἀποκρίνεσθε** indicate that the verb is second person plural and would be translated as "ye."
f. The verb endings **-ουσι(ν)** as in **ἀκούουσι(ν)**

and **-νται** and **-ντο** indicate that the verb is third person plural and would be translated as "they."

g. The present middle participle **λυόμενος**, **η**, **ον**, (loosing for himself), and the present passive participle **λυόμενος**, **η**, **ον**, (being loosed), are easily recognizable because of the presence of **μεν**. In the GNT the present passive participles are most often used with deponent verbs, which have active meanings.

h. The forms of the aorist passive indicative and the future passive indicative are easily recognizable. These two forms are arrived at by adding **θε** to the stem of the verb. Throughout the indicative, the **θε** is lengthened to **θη**. For example, the aorist passive indicative of **λύω** is **ἐλύθην**, which means "I was loosed," the Greek forms receiving the augment (or lengthening of the vowel) because it is a past tense. The future passive indicative is **λυθήσομαι**, means "I will be loosed."

4. Purchase a deck of 3 X 5 cards. Write the Greek word or phrase on one side of a card and the English translation on the opposite side. Review the cards repeatedly until you recognize the Greek terms and know their English meanings.
5. Another tool the student can employ to solidify a Greek word in his or her mind is to take a piece of paper and to write the word over and over until it is firmly fixed in the mind.

THIRTY-ONE

Greek Frequency Vocabulary

Nouns, verbs, adjectives and adverbs are listed in groups. Each group is divided into the general number of times each word occurs in the GNT.

NOUNS

Note: with few exceptions, most Greek nouns ending in -ος are masculine. Most nouns ending in -η are feminine, and all nouns ending in -μα and -ον are neuter.

English cognates are included in parentheses, where applicable.

MORE THAN 1300 TIMES:

θεός, god, God (theocracy)

. . .

MORE THAN 700 TIMES:

κύριος, lord, master, the Lord

MORE THAN 500 TIMES:

ἄνθρωπος, man, human (anthropology)

Χριστός, Christ

MORE THAN 400 TIMES:

πατήρ, father, Father (patriarch)

MORE THAN 300 TIMES:

ἡμέρα, day, lifetime, time period (ephemeral means short-lived)

πνεῦμα, wind, breath, spirit, Spirit (pneumatic)

υἱός, son

ἀδελφός brother (Philadelphia)

λόγος, (from λέγω, I say) word, speech, message, book, volume (logic)

MORE THAN 200 TIMES:

οὐρανός, heaven (Uranus)

μαθητής, disciple, student. Literally = learner

γῆ, earth, land (geography)

πίστις, (from πείθω, I believe) faith, belief, trust

ὄνομα, name

ἀνήρ, man, husband

γυνή, woman, wife (gynecology)

. . .

MORE THAN 100 TIMES:

νόμος, law, principle (nomology is the science or study of laws)

κόσμος, world, mankind, earth, decoration, adornment (cosmos)

χείρ, hand (chiropractor)

ἄγγελος, messenger, angel

ὄχλος, crowd, multitude

ἁμαρτία, (from ἁμαρτάνω, I sin) sin, sin offering (hamartiology is the study of the nature of sin)

ἔργον, work

δόξα, opinion, glory (doxology)

βασιλεία, (βασιλεύς [king]) kingdom (basilica)

πόλις, city, town (police)

ἔθνος, nation, people, Gentile (ethnic)

καρδία, heart (cardiac)

χάρις, grace, favor, gratitude, gift

σάρξ, flesh, meat, body, sinful nature (a sarcophagus is a limestone coffin. This Greek word means "flesh-eating.")

προφήτης, prophet

λαός, people

σῶμα, body (psychosomatic)

φωνή, voice, sound, speech (phone)

ζωή, (from ζάω, I live) life, living, way of life (zoology)

δοῦλος, slave, servant

αἰών, age, eternity (eon)

ἀρχιερεύς, high priest

θάνατος, (from θνῄσκω, I die)) death (thanatology is the study of death)

δύναμις, (from δύναμαι, I am able) power, strength, authority (dynamo)

ἀγάπη, love, kiss of peace

βασιλεύς, king (basilica)
ἐκκλησία, congregation, assembly, church (ecclesiastic)
οἶκος, house (economics)
ἀλήθεια, truth, truthfulness, faithfulness (Alice)
ὥρα, hour
ψυχή, soul (psyche)
ἐξουσία, authority, power
ὁδός, road, way (odometer)
ὀφθαλμός, (from ὁράω, I see) eye (ophthalmic)

MORE THAN 90 TIMES:

τέκνον, (from τίκτω, I give birth) (technology)
αἷμα, blood, bloodshed (hematology)
ἄρτος, bread, loaf
οἰκία, (synonym of οἶκος) house
πούς, foot
δικαιοσύνη, righteousness, justice
εἰρήνη, peace (irenic, Irene)
θάλασσα, (related to ἅλας [salt]) sea, lake
τόπος, place, position, opportunity, passage (topography)

MORE THAN 80 TIMES:

καιρός, time, season, opportunity
μήτηρ, mother (maternal)

MORE THAN 70 TIMES:

ἀπόστολος, apostle
στόμα, mouth (stomach)
σημεῖον, sign, miracle

εὐγγαέλιον, good news, gospel (evangelical)
πρόσωπον, (from πρός [to] + ὁράω [I see]) face
ὕδωρ, water (hydroelectric)
κεφαλή, head
φῶς,(φαίνω, I shine) light (photograph)
ἱερόν, temple
πῦρ, fire (pyre)

MORE THAN 60 TIMES:

ῥῆμα, word, thing
σάββατον, sabbath
ἐντολή, commandment
πλοῖον, boat, ship
καρπός, fruit
γραμματεύς, (from γράφω, I write) scribe (grammar)
δαιμόνιον, demon
ὄρος, mountain (orology is the study of mountains)
θέλημα, will, desire
θρόνος, throne, seat
νύξ, night

MORE THAN 50 TIMES:

ἱμάτιον, garment
διδάσκαλος, (from διδάσκω, I teach) teacher (didactic)
λίθος, stone (lithosphere)
χαρά, joy
συναγωγή, (from σύν [with] + ἄγω [I gather]) collection, gathering, synagogue
ἀρχή, beginning, first, principle, ruler, realm (archangel)
χρόνος, time (chronology)
ἐλπίς, hope

ἐπαγγελία, promise
παιδίον, child (pediatric)
σοφία, skill, wisdom, philosophy

MORE THAN 40 TIMES:

γλῶσσα, tongue, language, bar, ingot (glossolalia)

γραφή, (from γράφω, I write) written document, scripture (graphic)

παραβολή, proverb, poem, parable

ἔτος, year

χρεία, (from χράομαι, I need)) need, use, duty

κρίσις, (from κρίνω, I judge) judgment, decision (critic)

φόβος, fear, terror, reverence (phobia)

φυλακή, (from φυλάσσω, I guard) guard, watch, prison (prophylactic)

θηρίον, wild animal

σωτηρία, (σῴζω) salvation

θλῖψις, trouble, tribulation, oppression

ναός, temple, shrine, palace

γενεά, (from γίνομαι, I happen) generation

σπέρμα, (from σπείρω, I sow) seed (sperm)

ἀνάστασις, (from ἀνά [up] + ἵστημι [I stand]) standing up, rising, insurrection, resurrection

μέρος, part

τιμή, honor

MORE THAN 30 TIMES:

μνημεῖον, tomb, monument

τέλος ,end, goal, tribute

θύρα, door (door)

πρόβατον, (from πρό [ahead] + βαίνω, [I go]) sheep

ἐπιθυμία, desire, lust
ἄρχων, (ἄρχω) ruler
μαρτυρία, (μάρτυς) testimony, martyrdom
ἀγρός, field, farm, country (agriculture)
ὀργή, wrath, anger
οὖς, ear
περιτομή, circumcision
προσευχή, prayer
μάρτυς, witness, martyr
βιβλίον, (synonym of βίβλος) scroll, papyrus strip
διακονία, service, ministry (deacon)
μέλος, body part, musical part, melody
οἶνος, wine
διαθήκη, covenant, testament
ἥλιος, sun (helium)
ὑπομονή, endurance, staying
ἄνεμος, wind (anemometer)
ἱερεύς, priest
παρρησία, boldness
πλῆθος, multitude (plural)
ποτήριον, (from πίνω, I drink) cup
σκότος, darkness
φυλή, tribe

MORE THAN 20 TIMES:

ἀρνίον, (from ἀρήν, lamb) lamb, small lamb
διδαχή, (from διδάσκω, I teach) teaching
συνείδησις, conscience
γνῶσις, (from γινώσκω, I know) knowledge
διάκονος, servant, minister
μάχαιρα, dagger, short-sword
μισθός, wages, reward

παράκλησις, (from παρά [from alongside] + καλέω [I call]) encouragement

πάσχα, (πάσχω) passover, passover lamb

θυγάτηρ, daughter

θυσία, (from θύω, I sacrifice) sacrifice

μυστήριον, mystery, secret

χώρα, place, land, country

ἔλεος, mercy

κρίμα, (from κρίνω [I judge]) judgment, decree, decision (critic)

κώμη, village (commune)

σταυρός, (from σταυρόω, I crucify) cross

ἀδελφή, (related to ἀδελφός, brother) sister

στρατιώτης, soldier

χήρα, widow

ἀδικία, unrighteousness

δένδρον, tree (rhododendron is Greek for "red tree")

ἑορτή, festival, feast

νεφέλη, cloud

πορνεία, sexual immorality (pornography)

ἀκοή, (from ἀκούω, I hear) hearing, report, obedience

ἀσθένεια, weakness

ἀστήρ, star (asteroid)

ἐπιστολή, letter, epistle

νοῦς, mind, thought

παῖς, child, slave (pedagogy)

παρουσία, coming, presence

σωτήρ, (from σῴζω, I save) savior, Savior

ἀμπελών, (related to ἄμπελος, vineyard) vineyard

εἰκών, image (icon)

ζῷον, (from ζάω, I live) living being, animal

θυσιαστήριον, (from θύω, I sacrifice) altar

σκεῦος, object, vessel

χιλιάς, group of a thousand
κοιλία, stomach, womb
μετάνοια, repentance
πληγή, plague, blow, wound
πλοῦτος, riches (plutarchy)
διδασκαλία, (from διδάσκω, I teach) teaching
πειρασμός, temptation, test (pirate)
τελώνης, tax collector
χιλίαρχος, commander of a thousand, tribune

10 TIMES:

αἰτία, cause
ἀκροβυστία, uncircumcision
ἀργύριον, silver, money
γένος, (from γίνομαι, I happen) family, race, kind, offspring
γονεύς, (from γίνομαι, I happen) parent
ἑκατοντάρχης, (from ἑκατόν [a hundred] + ἄρχω [I rule]) commander of a hundred, centurion
ἐπίγνωσις, knowledge
ἡγεμών, governor, leader, chief
ἰχθύς, fish
ξύλον, wood, tree
σκηνή, tent, Tabernacle
ὑπηρέτης, assistant
βάπτισμα, (from βάπτω, I baptize) washing, baptism
γεωργός, (from γῆ [earth]+ ἔργον [field]) farmer
δῶρον, (related to δίδωμι, I give) gift
κοινωνία, (related to κοινός, common) fellowship, participation
κριτής (from κρίνω, I judge) judge (critic)
κτίσις (from κτίζω, I create) creation, creature

μαρτύριον, testimony, martyrdom

παράπτωμα, (from παρά [beside], πίπτω [I fall]) trespass, sin

προφητεία, (from πρό, [before] + φημί [I say]) prophecy

ἀποκάλυψις, revelation, the Apocalypse

ἀπώλεια, destruction, ruin

ἀριθμός, number (arithmetic)

βλασφημία, abusive speech, blasphemy

δέησις, entreaty, request, prayer

δεσμός, (from δέω, I bind) bond, chain, imprisonment

θυμός, anger, soul, spirit, sorrow, mind

κόπος, (from κόπτω, I labor) labor, trouble

μήν, month

οἰκοδομή, building

ποιμήν, shepherd

πόλεμος, war (polemic)

πυλών, gate, porch

στέφανος, crown

ἀνάγκη, necessity

ἄφεσις, forgiveness, release, remission

βρῶμα, food

θησαυρός, (from τίθημι, I place) treasure, treasury

ἵππος, horse (hippopotamus)

πλήρωμα, fullness

ποταμός, river (Hippopotamus is Greek for "river horse.")

ῥίζα, root, origin

ὑποκριτής, hypocrite

χάρισμα, gift

γάμος, marriage (monogamous)

δεῖπνον, dinner, supper

δέσμιος, (from δέω, I bind) bound, prisoner

δηνάριον, denarius

ἐργάτης, (from ἔργον, a work) worker

εὐλογία, (from εὖ [good] + λέγω [I say]) blessing
ζῆλος, (from ζέω, I am zealous) zeal, jealousy
κλέπτης, (from κλέπτω, I steal) thief (kleptomaniac)
λύπη, grief, pain
νυμφίος, bridegroom
πάθημα, (from πάσχω, I suffer) suffering
σκοτία, darkness
συκῆ, fig tree
σφραγίς, seal
τέρας, wonder, marvel, omen
τροφή, (from τρέφω, I eat) food, provisions
ἀνομία, (from α [against] + νόμος [law]) lawlessness
εὐσέβεια, godliness, piety, religion
εὐχαριστία, thanksgiving, thankfulness, eucharist
θεμέλιος, (from τίθημι [I put, place]) foundation
θρίξ, hair, thread
κληρονόμος, (from κληρόω [I inherit] + νόμος [law]) heir
λῃστής, thief
οἰκουμένη, world
ὀψία, evening
παρθένος, virgin
πέτρα, rock (petrified)
σκάνδαλον, offense, cause of sin
τράπεζα, table
τύπος, image, pattern, example, sign
ὑπακοή, (from ὑπό [under] + ἀκούω [I hear]) obedience
χόρτος, grass (chloroform)
ἄκανθα, thorns
γράμμα, (from γράφω, I write) character, letter (of alphabet), epistle (grammar)
διαλογισμός (from διά [through] + λέγω [I speak]) thought, opinion, discussion

ἐνιαυτός, year, period, cycle

κληρονομία, (from κληρόω [I inherit] + νόμος [law]) inheritance

λύχνος, lamp

μακροθυμία, patience, endurance

μέτρον, measurement, dimension, poetic meter

μύρον, ointment (myrrh)

ὄφις, snake

πετεινόν, bird

σεισμός, (from σείω, I shake) earthquake, shakedown (extortion)

σῖτος, grain

τάλαντον, talent

φύσις, nature

ἀναστροφή, way of life, lifestyle

δράκων, dragon, giant serpent

ἐλαία, olive, olive tree

ἐλεημοσύνη, pity, alms

ζύμη, yeast, leaven

θερισμός, harvest

καπνός, smoke

παιδίσκη, young woman, servant (female)

παράδοσι, (from παρά [beside] + δίδωμι, [I give]) deliverance, tradition

τρόπος, way, manner, customs, kind of life, deportment

ἀλέκτωρ, rooster, cock

ἀσκός, wineskin, leather bag

αὐλή, courtyard, court

βαπτιστής, (from βάπτω, I baptize) baptizer, Baptist

βῆμα, step, platform, judgment seat

βουλή, counsel, council, senate

βροντή, thunder

γέεννα, hell, Gehenna

γόνυ, knee
διάνοια, mind, thought, intention
δίκτυον, net, lattice
ἔθος, custom, habit
κάλαμος, reed
κράτος, power, might
λιμός, famine
λυχνία, lampstand, candlestick
ὀδούς, tooth
οἰκοδεσπότης, householder
ὅραμα, (from ὁράω, I see) vision, sight
ὅριον, border, coast, domain, limit
πόρνη, prostitute (pornography)
πρόθεσις,(πρό, τίθημι) setting forth, offering, purpose
πῶλος, colt
ῥάβδος, stick, staff, scepter
σπουδή, (from σπεύδω, I hasten) haste, speed, zeal, pursuit, exertion
φιάλη, bowl
χοῖρος, pig
χρυσίον, (synonym of χρυσός) gold, golden
ἀγορά, market
ἅλυσις, chain
ἀνατολή, east, sunrise
ἀπιστία, unbelief
βρῶσις, food, eating, rust
δόλος, deceit
δωρεά, (from δίδωμι, I give) gift, bounty, grant
εἴδωλον, idol
ἔλαιον, olive oil
ἐλευθερία, (related to ἐλεύθερος, free) freedom
ἔπαινος, praise
κακία, (related to κακός, [adjective, evil], evil, trouble

καταβολή, (from κατά [down] + βάλλω [I cast]) foundation, building, structure

καύχημα, (from καυχάομαι, I boast) boasting, pride

καύχησις, (from καυχάομαι, I boast) boasting, pride

κέρας, horn, (army) wing, flank

κλάδος, branch

κλῆρος, lot, portion, clergy

κλῆσις, (from καλέω, I call) calling

κράβαττος, bed, cot

λίμνη, lake

νεανίσκος, (related to νέος, young) young man

νόσος, disease

πηγή, water spring

πρᾶγμα, (from πράσσω, I do) thing, deed

πραΰτης, gentleness

σάλπιγξ, (from σαλπίζω, I sound the trumpet) trumpet

σπλάγχνον, entrails, heart

χιτών, tunic, shirt

ψευδοπροφήτης, false prophet

ἁγιασμός, (related to ἅγος, [adjective] holy) holiness (hagiography)

ᾅδης, Hades

ἀκαθαρσία, (from α [not], καθαρός [clean], adjective) uncleanness

ἀπολύτρωσις, (ἀπό [from], λύω [loose]) redemption, release

ἀσέλγεια, sensuality

ἀσπασμός, (from ἀσπάζομαι, I greet) greeting

βίβλος, scroll, book, record

βίος, life

δάκρυον, teardrop

δεσπότης, (from δέω, I bind) master, lord, slaveowner

δικαίωμα, ordinance, requirement, righteousness

διωγμός, (from διώκω, I persecute) persecution
ἰσχύς, strength, might
κοινωνός, partaker, partner, companion
οἰκονόμος, manager, steward
ὅρκος, oath
παντοκράτωρ, all mighty, almighty
παρεμβολή, barracks, encampment
περιστερά, dove, pigeon
πλάνη, error
πλεονεξία, greediness, lust
πόρνος, sexually immoral man, prostitute chaser
πύλη, gate
στρατηγός, captain, commander, chief magistrate
σύνδουλος, (from σύν [with] + δοῦλος [slave]) fellow slave
ὑπόδημα, sandal, shoe
χρηστότης, kindness
χρυσός, gold
χωρίον, place, field
ψεῦδος, (from ψεύδομαι, I lie) lie, falsehood
ψεύστης, (from ψεύδομαι, I lie) liar

VERBS

2,000 OR MORE TIMES:
εἰμί, (εἰμί, I am) to be, exist (sum)
λέγω, I say, speak, tell

MORE THAN 700 times
ἔχω, I have

. . .

MORE THAN 600 TIMES:

γίνομαι, I become, am, am born, am created (Genesis)
ἔρχομαι, I come, go

MORE THAN 500 TIMES:

ποιέω, I do, make, work (poet)

MORE THAN 400 TIMES:

ὁράω, I see, view, perceive (oracle)
ἀκούω, I hear, heed, obey (acoustics)
δίδωμι, I give

MORE THAN 300 TIMES:

οἶδα, I know

MORE THAN 200 TIMES:

λαλέω, I speak
λαμβάνω, I take, receive, choose
πιστεύω, I believe, trust
ἀποκρίνομαι, I answer, reply
γινώσκω, I know, come to know, recognize (knowledge)
ἐξέρχομαι, I come out, go out
δύναμαι, I am able (dynamo)
θέλω, I want, will

MORE THAN 100 TIMES:

εἰσέρχομαι, (from εἰς [into] + ἔρχομαι [I come]) I enter
γράφω, I write, inscribe, record (graphic)
εὑρίσκω, I find (Eureka!)
ἐσθίω, I eat
ἵστημι, I stand, place, put
πορεύομαι, I go
καλέω, I call, summon, invite
ἐγείρω, I raise, lift up
ἀγαπάω, (related to ἀγάπη [noun = love]), I love
ἀφίημι, I forgive, permit, free, neglect, abandon
ζάω, I live
ἀποστέλλω, I send
βλέπω, I see
βάλλω, I throw, put, cast (ballistic)
παραδίδωμι, I deliver, deliver over
μένω, I remain, abide, stay
ἀπέρχομαι, (ἀπό [from] + ἔρχομαι [I go]), I go away, depart
ζητέω, I seek, inquire
κρίνω, I judge
ἀποθνῄσκω, I die
παρακαλέω, (παρά [alongside] + καλέω [I call]) I urge, exhort, comfort
ἀνίστημι, (ἀνά [up] + ἵστημι [I stand]), I raise up, set up, restore, arise
σῴζω, I save
αἴρω, I take up
δεῖ, (related to δέω [I bind]), it is necessary
τίθημι, I put, make, appoint

MORE THAN 90 TIMES:

γεννάω, I am father of, bear, beget, engender

διδάσκω, I teach (didactic)

περιπατέω, I walk, walk around (peripatetic = traveling from place to place)

φοβέω, (related to φόβος [fear]), I fear (phobia)

κάθημαι, I sit

ἀκολουθέω, I follow

ἀπόλλυμι, I destroy, lose

πίπτω, I fall

MORE THAN 80 TIMES:

ἄρχω, I begin, (middle), I rule over

πληρόω, I fill, fulfill

προσέρχομαι, I come, approach

προσεύχομαι, I pray

ἀναβαίνω, I go up, rise up, advance

ἐκβάλλω, I cast out

καταβαίνω, I come down, go down

MORE THAN 70 TIMES:

πέμπω, I send

ὑπάγω, I go away

ἀνοίγω, I open, unlock, disclose

βαπτίζω, I baptize, wash, dip

μαρτυρέω, I bear witness, suffer martyrdom

ἀποκτείνω, I kill

χαίρω, I rejoice, hail

πίνω, I drink

MORE THAN 60 TIMES:

αἰτέω, I ask

τηρέω, I keep, I guard
ἄγω, I bring, lead, go, celebrate
ἀπολύω, I release, divorce
φέρω, I bring, carry, (passive, I am extant)
φημί, I say
ἐρωτάω, I ask, question
δοκέω, I think, suppose, seem
δοξάζω, I glorify, extol, venerate
κηρύσσω, I proclaim, preach

MORE THAN 50 TIMES:

προσκυνέω, I fall down, worship, do obeisance
ὑπάρχω, I am, exist, possess, have advantage
ἀσπάζομαι, I greet
συνάγω, I gather, bring together
θεωρέω, (synonym of θεάομαι), I see, observe
δέχομαι, I take, receive
ἐπερωτάω, I ask, inquire
κράζω, I call out
εὐαλλελίζω, I proclaim good news
πείθω, I persuade, believe, trust
σπείρω, I sow

MORE THAN 40 TIMES:

παραλαμβάνω, I take
φανερόω, (synonym of φαίνω), I make known, reveal
ἀποδίδωμι, I give back, pay
κρατέω, I grasp, am strong, take possession
προσφέρω, I bring, offer
καθίζω, I sit, set, place
σταυρόω, I crucify

διώκω, I pursue, persecute
ἐπιγινώσκω, I know, look upon, witness, recognize, decide
κατοικέω, I settle, dwell, inhabit
ἁμαρτάνω, I sin
δέω, I bind, stop
διέρχομαι, I pass through
θαυμάζω, I marvel, wonder
θεραπεύω, I heal
φωνέω, (related to φωνή, [voice]), I call, shout
λύω, I loosen, release
πάσχω, I suffer
ἐργάζομαι, (related to ἔργον, work) I work
εὐλογέω, (εὖ [good] + [λέγω] I say) I bless
παρίστημι, I present, stand by

MORE THAN 30 TIMES:

ἑτοιμάζω, I prepare
κλαίω, I weep, cry
λογίζομαι, I count, think, calculate
μισέω, I hate
οἰκοδομέω, I build
ἅπτω, I touch, hold, grasp, light, ignite
δικαιόω, I pronounce righteous, justify
ἐπιτίθημι, I lay on, place, put, add
περισσεύω, I abound
πλανάω (πλάνη), I deceive
πράσσω, I do, accomplish
εὐχαριστέω, I give thanks
πειράζω, I tempt, test, try
ὑποτάσσω, I subject, submit
βούλομαι, I will, want
διακονέω, I serve, wait on

καυχάομαι, I boast, glory
παραγίνομαι, I come, appear
ἐπιστρέφω, I turn back, return, turn
ὀφείλω, I am obligated
ὑποστρέφω, I return
βλασφημέω, I verbally abuse, blaspheme
μετανοέω, I repent
ἀρνέομαι, I deny
ἀσθενέω, I am weak
ἐκπορεύομαι, I go, come out
ἀναγινώσκω, I read
ἐλπίζω, I hope
ἔξεστιν, it is right, possible
καθαρίζω, I cleanse
φαίνω, I appear, shine, cause to shine
φυλάσσω, I guard, keep

MORE THAN 20 TIMES:

ἀγοράζω, I buy
δείκνυμι, I show
ἐπικαλέω, I call on
συνέρχομαι, I come together, go together
ἐπιτιμάω, I rebuke
παρέρχομαι, I pass by, pass away
προσκαλέω, I call, summon
σκανδαλίζω, I cause somebody to sin
φεύγω, I flee
ἁγιάζω, I sanctify
ἀδικέω, I harm
γαμέω, I marry
ἐλεέω, I show mercy
ἡγέομαι, I lead, consider, count, regard

ἰσχύω, I am able, strong
νικάω, I conquer (Nike was the Greek goddess of victory)
προφητεύω, I prophesy
τελέω, I finish
βαστάζω, I bear
ἐνδύω, I wear, put on
καταργέω, I leave idle, occupy, nullify
ἀποκαλύπτω, I reveal
ἥκω, I have come, am present
ἰάομαι, I heal
λυπέω, I grieve, pain
ὀμνύω, I swear
ὁμολογέω, I confess, praise
φρονέω, I am wise, think
ἀναβλέπω, I receive sight
γνωρίζω, I make known
δουλεύω, I serve as a slave
κελεύω, I command
μανθάνω, I learn
φιλέω, I love, have affection for, kiss
ἀναιρέω, I destroy, kill
καταλείπω, I leave, forsake
κεῖμαι, I lie down, am valid for
πάρειμι, I am present
πίμπλημι, I fill, fulfill
προσέχω, I pay attention to
ἀνάγω, I take up, raise, offer up, bring
κατηγορέω, I accuse, reproach
κοπιάω, I toil, grow tired, am weary
κωλύω, I hinder
μιμνῄσκομαι, I remember, remind
πεινάω, I am hungry
τελειόω, I finish, make perfect

χαρίζομαι, I give, favor, forgive
ἀγνοέω, I am ignorant, ignore
γρηγορέω, I am or become alert, awake
δέομαι, I pray, ask, beg
δοκιμάζω, I prove, approve, test
ἐκλέγομαι, I choose
θεάομαι, I see
καθεύδω, I sleep
κατεργάζομαι, I work, prepare, make, oppress
πωλέω, I sell
αὐξάνω, I grow, cause to grow
βασιλεύω, I reign, make a king
ἐνεργέω, I work
εὐδοκέω, I am well pleased, consent, enjoy
θερίζω, I reap
καθίστημι, I put down, bring to, put in charge, make
λατρεύω, I serve, worship
μνημονεύω (μιμνῄσκομαι), I remember, mention
στρέφω, I turn
τιμάω, I honor
ὑπακούω, I obey, hear, listen
νηστεύω, I fast
προάγω, I go before, elevate
τρέχω, I run
ὑψόω, I lift up, I exalt

MORE THAN 10 TIMES:

ἀπέχω, I receive, obtain, get, be far off
διακρίνω, I evaluate, consider, doubt
ἐπαίρω, I lift up
ἐπιλαμβάνομαι, I take hold of
μεριμνάω, I am anxious for

παρατίθημι, I set before, commend

εἰσπορεύομαι, I enter, go in

ἐπιβάλλω, I lay hands on, throw

ἐπιτρέπω, I entrust to, permit, allow

κατακρίνω, I condemn

κληρονομέω, I inherit

κοιμάω, I sleep, I am calm

κρύπτω, I hide

προστίθημι, I put, add

τίκτω, I bear children

ἀρέσκω, I please

ἐλέγχω, I reprove, convict

ἐξίστημι, I amaze, confuse

καταλύω, I destroy, lodge

κατέχω, I hold, withhold

κερδαίνω, I gain

νίπτω, I wash

περιτέμνω, I circumcise

ῥύομαι, I deliver

ταράσσω, I trouble

ὑπομένω, I endure, remain, wait upon

ἀθετέω, I reject, rebel against

ἀνακρίνω, I examine

διαλογίζομαι, I reason, discuss, consider

διατάσσω, I assign, arrange, command

διψάω, I thirst

ἐκτείνω, I stretch out

ἐκχέω, I pour out, bring forth, come forth

ἐμβαίνω, I embark, enter

ἐπιθυμέω, I desire

ἐπιμένω, remain, stay, continue

κατέρχομαι, I come down, go down, return

κλείω, I close, shut

παρέχω, I provide, cause, grant
προσδοκάω, I wait for, expect
συνλαμβάνω, I seize
συνίστημι, I associate with, recommend, unite, collect
τολμάω, I dare, act rashly
ὑστερέω, I lack, am late, postpone
χορτάζω, I feed, fill
ἀνέχω, I lift, esteem, hinder, stop, bear
ἀπάγω, I lead away
γεύομαι, I taste, experience
δέρω, I beat, skin, flay
διαμαρτύρομαι, testify, declare solemnly
ἐντέλλω, I command
καταλαμβάνω, I take, overtake, reach
κτίζω, I create, build, found
μοιχεύω, I commit adultery
νομίζω, I think, suppose
ξηραίνω, I dry up
ὁμοιόω, I make like
παύω, I cease, stop
ποτίζω, I give a drink
προλέγω, I say ahead of time, foretell
σαλεύω, I shake
συμφέρω, I am better, bring together
σφραγίζω, I seal
ὠφελέω, I gain, benefit
ἀνάκειμαι, I sit at table, recline at table
ἀναχωρέω, I go back, depart
ἀνθίστημι, I resist, oppose
ἀπειθέω, I disbelieve, disobey
ἁρπάζω, I snatch away
ἀτενίζω, I look steadfastly
ἀφίστημι, I withdraw, remove, depart, leave

ἐπίσταμαι, I know
εὐφραίνω, I cheer, I rejoice
θύω, I sacrifice, murder
κατανοέω, I understand, consider
κατεσθίω, I devour
κλάω, I break
κοινόω, I defile
μερίζω, I divide, distribute
νοέω, I understand
προσδέχομαι, I receive, wait for
απεινόω, I humble, humiliate
ἀναλαμβάνω, I take up, raise, undertake
δαιμονίζομαι, I am demon possessed, demonized
διαλέγομαι, I dispute, discuss
διαφέρω, I am better, spread, carry over
ἐκπλήσσω, I am amazed
ἐμπαίζω, I mock

ADJECTIVES

MORE THAN 70 TIMES:

πονηρός, wicked, evil
ἐμός, (from ἐγώ, I) my
δώδεκα, twelve
αἰώνιος, eternal

MORE THAN 60 TIMES:

τρεῖς, three
πιστός, faithful, reliable, trustworthy
πρεσβύτερος, older, elder

ἀγαπητός, beloved
δεξιός, right, right hand
μέσος, middle, (prep +gen) in the middle, (adverb) among

MORE THAN 40 TIMES:

ἔσχατος, last
κακός, evil, bad, wrong, harm
μακάριος, blessed, happy, fortunate
τυφλός, blind
ἔρημος, desolate, wilderness, desert
ἁμαρτωλός, sinner, sinful
μικρός, small, little
ὅμοιος, like, similar
μόνος, only, alone
καινός, new
ἄξιος, worthy
τέσσαρες, four
λοιπός, rest, remaining, other

MORE THAN 30 TIMES:

πέντε, five
διάβολος, enemy, adversary, devil
ἅπας, all, every
πτωχός, poor
ὀλίγος, little, few
ἀκάθαρτος, unclean
ἐχθρός, hostile, enemy
δυνατός, possible, strong, able

MORE THAN 20 TIMES:

ἰσχυρός, strong, mighty
φίλος, friend, beloved, friendly, dear, pleasant, welcome
ἀληθινός, true, genuine
καθαρός, pure
πλούσιος, rich
ἀληθής, true, honest, genuine
ἀσθενής, weak, weakness
πνευματικός, spiritual
δέκα, ten
λευκός, white
σός, your (singular)
νέος, new, young (a neologism is a newly coined word)
ἄπιστος, incredible, unbelievable, unbelieving
ἐλεύθερος, free
ἐκλεκτός, elect, chosen
τεσσεράκοντα, forty
10 or more times:
σοφός, wise, skilled
ἐπουράνιος, heavenly
παλαιός, old (paleontology)
τέλειος, complete, perfect
κρείττων, better
φανερός, known, plain
χρυσοῦς, golden
ἑκατόν, hundred
ἕτοιμος, ready
κενός, empty, foolish, worthless (cenotaph means "empty tomb" in Greek)
κρυπτός, hidden
πλήρης, full
γνωστός, known, friend
γυμνός, naked
ἀμφότεροι, both, all

ἐλάχιστος, least
κοινός, common, common assembly
κωφός, deaf, unable to speak
νήπιος, child (nipple)
ξένος, stranger, strange (xenophobia)
περισσός, more, remaining, excessive, to the full
φρόνιμος, wise
χωλός, crippled
ἕξ, six
συνεργός, fellow worker
τίμιος, precious, rare
ὕψιστος, highest, most high, Most High
ἄδικος, unjust, unrighteous
μωρός, foolish (moron)
Ῥωμαῖος, Roman
ἄφρων, fool
Γαλιλαῖος, Galilean
εἴκοσι, twenty
τριάκοντα, thirty
ὑγιής, whole, healthy
ὑμέτερος, your (plural)
ὑψηλός, high, proud
χείρων, worse
χίλιοι, thousand
ἀδύνατος, unable, impossible, powerless, indissoluble
ἀρχαῖος, old, ancient
ἔνοχος, liable, guilty
κλητός, called, invited, guest
παραλυτικός, paralytic
ποικίλος, many colored, various kinds

ADVERBS

. . .

MORE THAN 800 TIMES:

καί, and, also, even, and yet, but

MORE THAN 200 TIMES:

οὕτως, thus, so, in this way

MORE THAN 100 TIMES:

τότε, then

νῦν, now, present

πάλιν, again, in so far as

ἐκεῖ, there

MORE THAN 90 TIMES:

ἔτι, yet, still, again

70 TIMES:

μόνος, only, alone

MORE THAN 60 TIMES:

ἤδη, now, already

ὧδε, here, in this way, so, thus

MORE THAN 50 TIMES:

πρῶτος, first, prominent, former

εὐθύς, immediately, next, suddenly

. . .

MORE THAN 40 TIMES:

οὐκέτι, no longer
οὐδέ, and not, neither, nor, not even
ἔξω, (+gen) out, outside, outsider
πάντοτε, always
σήμερον, today

MORE THAN 30 TIMES:

ἐκεῖθεν, from there
καλῶς, well
ἄρτι, now
εὐθέως, immediately, at once, suddenly
ὁμοίως, likewise

MORE THAN 20 TIMES:

οὔπω, not yet
οὗ, where, to where
μηκέτι, no longer, no more
νυνί, (synonym of νῦν) now
πολύς, much, many
10 or more times:
ἀληθῶς, truly
παραχρῆμα, immediately
ἐπαύριον, tomorrow
ὡσαύτως, likewise
ἔπειτα, then
κακῶς, badly, severely, wrong, fierce
οὐδέποτε, never
πλησίον, (+gen) near, neighbor

εἶτα, then, next
λοιπός, rest, remaining, other
ταχέως, quickly, soon
ἅπαξ, once, once for all
αὔριον, tomorrow
μακρόθεν, far off, from afar
ἄνωθεν, from above, again
ἔσωθεν, within, inside, from within
λίαν, exceedingly
μάλιστα, especially, exceedingly
περισσοτέρως, even more so
πρωΐ, in the morning
ταχύς, swift, quickly, soon
τρίς, three times
ἔξωθεν, (+gen) from outside, outside
ὁμοθυμαδόν, of one accord
σφόδρα, very much
ὕστερος, afterwards, last, later
ἐντεῦθεν, from here, on this side. . . on that side
μακράν, far, far off, far away
ὄντως, really, indeed
πρότερος, former, formerly

CONJUNCTIONS

MORE THAN 8,000 TIMES:

καί, and, also, even, and yet, but

MORE THAN 2,000 TIMES:

δέ, but, and, now

. . .

MORE THAN 1,200 TIMES:

ὅτι, that, because, since

MORE THAN 1,000 TIMES:

γάρ, for, since, then, indeed

MORE THAN 600 TIMES:

ἵνα, that, in order that, so that

MORE THAN 400 TIMES:

οὖν, therefore, thus

εἰ, if, whether

MORE THAN 200 TIMES:

ἐάν, if

τέ, and, so, both/and

MORE THAN 100 TIMES:

ὡς, as, like, while, that

καθώς, as, just as, inasmuch as

ὅταν, when, whenever

ὅτε, when, while

MORE THAN 90 TIMES:

οὐδέ, and not, neither, nor, not even

. . .

MORE THAN 80 TIMES:

οὔτε, neither
ὥστε, so that, in order that, thus
ὅπου, where, whereas

MORE THAN 60 TIMES:

εἴτε, or, either/or, even if

MORE THAN 50 TIMES:

διό, therefore
ὅπως, so that, in order that, how that

MORE THAN 40 TIMES:

μηδέ, nor, and not
ἄρα, then, therefore
μή, not, lest

MORE THAN 30 TIMES:

ἕως, until, when, up to, as far as
μήτε, and not, neither . . . nor

MORE THAN 20 TIMES:

ὥσπερ, as, just as
πλήν, (+gen) only, except (prep.), but, nevertheless (conj.)
πόθεν, from where?
ἐπεί, since, because, when

διότι, because, for, therefore
μήποτε, lest, perhaps

13 TIMES OR LESS:
καθάπερ, as, just as, like
πρίν, before, formerly
ἐπειδή, since, because, for

GREEK TO ENGLISH LEXICON

α

ἁμαρτάνω, I sin

ἁμαρτία, (from ἁμαρτάνω, I sin) sin, sin offering (hamartiology is the study of the nature of sin)

ἁμαρτωλός, sinner, sinful

ἀμπελών, (related to ἄμπελος, vineyard) vineyard

ἀμφότεροι, both, all

ἀγαπάω, (related to ἀγάπη [noun = love]), I love

ἀγάπη, love, kiss of peace

ἀγαπητός, beloved

ἄγγελος, messenger, angel

ἁγιάζω, I sanctify

ἁγιασμός, (related to ἅγος, [adjective] holy) holiness (hagiography)

ἀγνοέω, I am ignorant, ignore

ἀγορά, market

ἀγοράζω, I buy

ἀγρός, field, farm, country (agriculture)
ἄγω, I bring, lead, go, celebrate
ἀδελφή, (related to ἀδελφός, brother) sister
ἀδελφός brother (Philadelphia)
ᾅδης, Hades
ἀδικέω, I harm
ἀδικία, unrighteousness
ἄδικος, unjust, unrighteous
ἀδύνατος, unable, impossible, powerless, indissoluble
ἀθετέω, I reject, rebel against
αἷμα, blood, bloodshed (hematology)
αἴρω, I take up
αἰτέω, I ask
αἰτία, cause
αἰών, age, eternity (eon)
αἰώνιος, eternal
ἀκαθαρσία, (from α [not], καθαρός [clean], adjective) uncleanness
ἀκάθαρτος, unclean
ἄκανθα, thorns
ἀκοή, (from ἀκούω, I hear) hearing, report, obedience
ἀκολουθέω, I follow
ἀκούω, I hear, heed, obey (acoustics)
ἀκροβυστία, uncircumcision
ἀλέκτωρ, rooster, cock
ἀλήθεια, truth, truthfulness, faithfulness (Alice)
ἀληθής, true, honest, genuine
ἀληθινός, true, genuine
ἀληθῶς, truly
ἅλυσις, chain
ἀναβαίνω, I go up, rise up, advance
ἀναβλέπω, I receive sight

ἀναγινώσκω, I read

ἀνάγκη, necessity

ἀνάγω, I take up, raise, offer up, bring

ἀναιρέω, I destroy, kill

ἀνάκειμαι, I sit at table, recline at table

ἀνακρίνω, I examine

ἀναλαμβάνω, I take up, raise, undertake

ἀνάστασις, (from ἀνά [up] + ἵστημι [I stand]) standing up, rising, insurrection, resurrection

ἀναστροφή, way of life, lifestyle

ἀνατολή, east, sunrise

ἀναχωρέω, I go back, depart

ἄνεμος, wind (anemometer)

ἀνέχω, I lift, esteem, hinder, stop, bear

ἀνήρ, man, husband

ἀνθίστημι, I resist, oppose

ἄνθρωπος, man, human (anthropology)

ἀνίστημι, (ἀνά [up] + ἵστημι [I stand]), I raise up, set up, restore, arise

ἀνομία, (from α [against] + νόμος [law]) lawlessness

ἀνοίγω, I open, unlock, disclose

ἄνωθεν, from above, again

ἄξιος, worthy

ἀπάγω, I lead away

ἅπαξ, once, once for all

ἅπας, all, every

ἀπειθέω, I disbelieve, disobey

απεινόω, I humble, humiliate

ἀπέρχομαι, (ἀπό [from] + ἔρχομαι [I go]), I go away, depart

ἀπέχω, I receive, obtain, get, be far off

ἀπιστία, unbelief

ἄπιστος, incredible, unbelievable, unbelieving

ἀποδίδωμι, I give back, pay

ἀποθνῄσκω, I die

ἀποκαλύπτω, I reveal

ἀποκάλυψις, revelation, the Apocalypse

ἀποκρίνομαι, I answer, reply

ἀποκτείνω, I kill

ἀπόλλυμι, I destroy, lose

ἀπολύτρωσις, (ἀπό [from], λύω [loose]) redemption, release

ἀπολύω, I release, divorce

ἀποστέλλω, I send

ἀπόστολος, apostle

ἅπτω, I touch, hold, grasp, light, ignite

ἀπώλεια, destruction, ruin

ἄρα, then, therefore

ἀργύριον, silver, money

ἀρέσκω, I please

ἀριθμός, number (arithmetic)

ἀρνέομαι, I deny

ἀρνίον, (from ἀρήν, lamb) lamb, small lamb

ἁρπάζω, I snatch away

ἄρτι, now

ἄρτος, bread, loaf

ἀρχαῖος, old, ancient

ἀρχή, beginning, first, principle, ruler, realm (archangel)

ἀρχιερεύς, high priest

ἄρχω, I begin, (middle), I rule over

ἄρχων, (ἄρχω) ruler

ἀσέλγεια, sensuality

ἀσθένεια, weakness

ἀσθενέω, I am weak

ἀσθενής, weak, weakness
ἀσκός, wineskin, leather bag
ἀσπάζομαι, I greet
ἀσπασμός, (from ἀσπάζομαι, I greet) greeting
ἀστήρ, star (asteroid)
ἀτενίζω, I look steadfastly
αὐλή, courtyard, court
αὐξάνω, I grow, cause to grow
αὔριον, tomorrow
ἄφεσις, forgiveness, release, remission
ἀφίημι, I forgive, permit, free, neglect, abandon
ἀφίστημι, I withdraw, remove, depart, leave
ἄφρων, fool

β

βάλλω, I throw, put, cast (ballistic)
βαπτίζω, I baptize, wash, dip
βάπτισμα, (from βάπτω, I baptize) washing, baptism
βαπτιστής, (from βάπτω, I baptize) baptizer, Baptist
βασιλεία, (βασιλεύς [king]) kingdom (basilica)
βασιλεύς, king (basilica)
βασιλεύω, I reign, make a king
βαστάζω, I bear
βῆμα, step, platform, judgment seat
βιβλίον, (synonym of βίβλος) scroll, papyrus strip
βίβλος, scroll, book, record
βίος, life
βλασφημέω, I verbally abuse, blaspheme
βλασφημία, abusive speech, blasphemy
βλέπω, I see
βουλή, counsel, council, senate
βούλομαι, I will, want

βροντή, thunder
βρῶμα, food
βρῶσις, food, eating, rust

γ

γαμέω, I marry
γάμος, marriage (monogamous)
Γαλιλαῖος, Galilean
γάρ, for, since, then, indeed
γέεννα, hell, Gehenna
γενεά, (from γίνομαι, I happen) generation
γεννάω, I am father of, bear, beget, engender
γένος, (from γίνομαι, I happen) family, race, kind, offspring
γεύομαι, I taste, experience
γεωργός, (from γῆ [earth]+ ἔργον [field]) farmer
γῆ, earth, land (geography)
γίνομαι, I become, am, am born, am created (Genesis)
γινώσκω, I know, come to know, recognize (knowledge)
γλῶσσα, tongue, language, bar, ingot (glossolalia)
γνωρίζω, I make known
γνῶσις, (from γινώσκω, I know) knowledge
γνωστός, known, friend
γονεύς, (from γίνομαι, I happen) parent
γόνυ, knee
γράμμα, (from γράφω, I write) character, letter (of alphabet), epistle (grammar)
γραμματεύς, (from γράφω, I write) scribe (grammar)
γραφή, (from γράφω, I write) written document, scripture (graphic)
γράφω, I write, inscribe, record (graphic)
γρηγορέω, I am or become alert, awake
γυμνός, naked

γυνή, woman, wife (gynecology)

δ

δαιμονίζομαι, I am demon possessed, demonized
δαιμόνιον, demon
δάκρυον, teardrop
δέ, but, and, now
δέησις, entreaty, request, prayer
δεῖ, (related to δέω [I bind]), it is necessary
δείκνυμι, I show
δεῖπνον, dinner, supper
δέκα, ten
δένδρον, tree (rhododendron is Greek for "red tree")
δεξιός, right, right hand
δέομαι, I pray, ask, beg
δέρω, I beat, skin, flay
δέσμιος, (from δέω, I bind) bound, prisoner
δεσμός, (from δέω, I bind) bond, chain, imprisonment
δεσπότης, (from δέω, I bind) master, lord, slaveowner
δέχομαι, I take, receive
δέω, I bind, stop
δηνάριον, denarius
διαμαρτύρομαι, testify, declare solemnly
διάβολος, enemy, adversary, devil
διαθήκη, covenant, testament
διακονέω, I serve, wait on
διακονία, service, ministry (deacon)
διάκονος, servant, minister
διακρίνω, I evaluate, consider, doubt
διαλέγομαι, I dispute, discuss
διαλογίζομαι, I reason, discuss, consider
διαλογισμός (from διά [through] + λέγω [I speak]) thought, opinion, discussion

διάνοια, mind, thought, intention
διατάσσω, I assign, arrange, command
διαφέρω, I am better, spread, carry over
διδασκαλία, (from διδάσκω, I teach) teaching
διδάσκαλος, (from διδάσκω, I teach) teacher (didactic)
διδάσκω, I teach (didactic)
διδαχή, (from διδάσκω, I teach) teaching
δίδωμι, I give
διέρχομαι, I pass through
δικαιοσύνη, righteousness, justice
δικαιόω, I pronounce righteous, justify
δικαίωμα, ordinance, requirement, righteousness
δίκτυον, net, lattice
διό, therefore
διότι, because, for, therefore
διψάω, I thirst
διωγμός, (from διώκω, I persecute) persecution
διώκω, I pursue, persecute
δοκέω, I think, suppose, seem
δοκιμάζω, I prove, approve, test
δόλος, deceit
δόξα, opinion, glory (doxology)
δοξάζω, I glorify, extol, venerate
δουλεύω, I serve as a slave
δοῦλος, slave, servant
δράκων, dragon, giant serpent
δύναμαι, I am able (dynamo)
δύναμις, (from δύναμαι, I am able) power, strength, authority (dynamo)
δυνατός, possible, strong, able
δώδεκα, twelve
δωρεά, (from δίδωμι, I give) gift, bounty, grant
δῶρον, (related to δίδωμι, I give) gift

ε

ἐμβαίνω, I embark, enter
ἐμός, (from ἐγώ, I) my
ἐμπαίζω, I mock
ἐάν, if
ἐγείρω, I raise, lift up
ἔθνος, nation, people, Gentile (ethnic)
ἔθος, custom, habit
εἰ, if, whether
εἰμί, (εἰμί, I am) to be, exist (sum)
εἴδωλον, idol
εἴκοσι, twenty
εἰκών, image (icon)
εἰρήνη, peace (irenic, Irene)
εἰσέρχομαι, (from εἰς [into] + ἔρχομαι [I come]) I enter
εἰσπορεύομαι, I enter, go in
εἶτα, then, next
εἴτε, or, either/or, even if
ἑκατόν, hundred
ἑκατοντάρχης, (from ἑκατόν [a hundred] + ἄρχω [I rule]) commander of a hundred, centurion
ἐκβάλλω, I cast out
ἐκεῖ, there
ἐκεῖθεν, from there
ἐκκλησία, congregation, assembly, church (ecclesiastic)
ἐκλέγομαι, I choose
ἐκλεκτός, elect, chosen
ἐκπλήσσω, I am amazed
ἐκπορεύομαι, I go, come out
ἐκτείνω, I stretch out
ἐκχέω, I pour out, bring forth, come forth
ἐλαία, olive, olive tree
ἔλαιον, olive oil

ἐλάχιστος, least
ἐλέγχω, I reprove, convict
ἐλεέω, I show mercy
ἐλεημοσύνη, pity, alms
ἔλεος, mercy
ἐλευθερία, (related to ἐλεύθερος, free) freedom
ἐλεύθερος, free
ἐλπίζω, I hope
ἐλπίς, hope
ἐνδύω, I wear, put on
ἐνεργέω, I work
ἐνιαυτός, year, period, cycle
ἔνοχος, liable, guilty
ἐντέλλω, I command
ἐντεῦθεν, from here, on this side. . . on that side
ἐντολή, commandment
ἕξ, six
ἐξέρχομαι, I come out, go out
ἔξεστιν, it is right, possible
ἐξίστημι, I amaze, confuse
ἐξουσία, authority, power
ἔξω, (+gen) out, outside, outsider
ἔξωθεν, (+gen) from outside, outside
ἑορτή, festival, feast
ἐπαγγελία, promise
ἔπαινος, praise
ἐπαίρω, I lift up
ἐπαύριον, tomorrow
ἐπεί, since, because, when
ἐπειδή, since, because, for
ἔπειτα, then
ἐπερωτάω, I ask, inquire
ἐπιμένω, remain, stay, continue

ἐπιβάλλω, I lay hands on, throw
ἐπιγινώσκω, I know, look upon, witness, recognize, decide
ἐπίγνωσις, knowledge
ἐπιθυμέω, I desire
ἐπιθυμία, desire, lust
ἐπικαλέω, I call on
ἐπιλαμβάνομαι, I take hold of
ἐπίσταμαι, I know
ἐπιστολή, letter, epistle
ἐπιστρέφω, I turn back, return, turn
ἐπιτιμάω, I rebuke
ἐπιτίθημι, I lay on, place, put, add
ἐπιτρέπω, I entrust to, permit, allow
ἐπουράνιος, heavenly
ἐργάζομαι, (related to ἔργον, work) I work
ἐργάτης, (from ἔργον, a work) worker
ἔργον, work
ἔρημος, desolate, wilderness, desert
ἔρχομαι, I come, go
ἐρωτάω, I ask, question
ἐσθίω, I eat
ἔσχατος, last
ἔσωθεν, within, inside, from within
ἔτι, yet, still, again
ἑτοιμάζω, I prepare
ἕτοιμος, ready
ἔτος, year
εὐαλλελίζω, I proclaim good news
εὐγγαέλιον, good news, gospel (evangelical)
εὐδοκέω, I am well pleased, consent, enjoy
εὐθέως, immediately, at once, suddenly
εὐθύς, immediately, next, suddenly
εὐλογέω, (εὖ [good] + [λέγω] I say) I bless

εὐλογία, (from εὖ [good] + λέγω [I say]) blessing
εὑρίσκω, I find (Eureka!)
εὐσέβεια, godliness, piety, religion
εὐφραίνω, I cheer, I rejoice
εὐχαριστέω, I give thanks
εὐχαριστία, thanksgiving, thankfulness, eucharist
ἐχθρός, hostile, enemy
ἔχω, I have
ἕως, until, when, up to, as far as

ζ

ζάω, I live
ζῆλος, (from ζέω, I am zealous) zeal, jealousy
ζητέω, I seek, inquire
ζύμη, yeast, leaven
ζωή, (from ζάω, I live) life, living, way of life (zoology)
ζῷον, (from ζάω, I live) living being, animal

η

ἡμέρα, day, lifetime, time period (ephemeral means short-lived)
ἡγεμών, governor, leader, chief
ἡγέομαι, I lead, consider, count, regard
ἤδη, now, already
ἥκω, I have come, am present
ἥλιος, sun (helium)

θ

θάλασσα, (related to ἅλας [salt]) sea, lake
θάνατος, (from θνῄσκω, I die)) death (thanatology is the study of death)
θαυμάζω, I marvel, wonder
θεμέλιος, (from τίθημι [I put, place]) foundation

θεάομαι, I see
θέλημα, will, desire
θέλω, I want, will
θεός, god, God (theocracy)
θεραπεύω, I heal
θερίζω, I reap
θερισμός, harvest
θεωρέω, (synonym of θεάομαι), I see, observe
θηρίον, wild animal
θησαυρός, (from τίθημι, I place) treasure, treasury
θλῖψις, trouble, tribulation, oppression
θρίξ, hair, thread
θρόνος, throne, seat
θυμός, anger, soul, spirit, sorrow, mind
θυγάτηρ, daughter
θύρα, door (door)
θυσία, (from θύω, I sacrifice) sacrifice
θυσιαστήριον, (from θύω, I sacrifice) altar
θύω, I sacrifice, murder

ι

ἱμάτιον, garment
ἰάομαι, I heal
ἱερεύς, priest
ἱερόν, temple
ἵνα, that, in order that, so that
ἵππος, horse (hippopotamus)
ἵστημι, I stand, place, put
ἰσχυρός, strong, mighty
ἰσχύς, strength, might
ἰσχύω, I am able, strong
ἰχθύς, fish

κ

καθάπερ, as, just as, like
καθαρίζω, I cleanse
καθαρός, pure
καθεύδω, I sleep
κάθημαι, I sit
καθίζω, I sit, set, place
καθίστημι, I put down, bring to, put in charge, make
καθώς, as, just as, inasmuch as
καί, and, also, even, and yet, but
καί, and, also, even, and yet, but
καινός, new
καιρός, time, season, opportunity
κακία, (related to κακός, [adjective, evil], evil, trouble
κακός, evil, bad, wrong, harm
κακῶς, badly, severely, wrong, fierce
κάλαμος, reed
καλέω, I call, summon, invite
καλῶς, well
καπνός, smoke
καρδία, heart (cardiac)
καρπός, fruit
καταβαίνω, I come down, go down
καταβολή, (from κατά [down] + βάλλω [I cast]) foundation, building, structure
κατακρίνω, I condemn
καταλαμβάνω, I take, overtake, reach
καταλείπω, I leave, forsake
καταλύω, I destroy, lodge
κατανοέω, I understand, consider
καταργέω, I leave idle, occupy, nullify
κατεργάζομαι, I work, prepare, make, oppress
κατέρχομαι, I come down, go down, return

κατεσθίω, I devour

κατέχω, I hold, withhold

κατηγορέω, I accuse, reproach

κατοικέω, I settle, dwell, inhabit

καυχάομαι, I boast, glory

καύχημα, (from καυχάομαι, I boast) boasting, pride

καύχησις, (from καυχάομαι, I boast) boasting, pride

κεῖμαι, I lie down, am valid for

κελεύω, I command

κενός, empty, foolish, worthless (cenotaph means "empty tomb" in Greek)

κέρας, horn, (army) wing, flank

κερδαίνω, I gain

κεφαλή, head (microcephalic is Greek for "small head")

κηρύσσω, I proclaim, preach

κλάδος, branch

κλαίω, I weep, cry

κλάω, I break

κλείω, I close, shut

κλέπτης, (from κλέπτω, I steal) thief (kleptomaniac)

κληρονομέω, I inherit

κληρονομία, (from κληρόω [I inherit] + νόμος [law]) inheritance

κληρονόμος, (from κληρόω [I inherit] + νόμος [law]) heir

κλῆρος, lot, portion, clergy

κλῆσις, (from καλέω, I call) calling

κλητός, called, invited, guest

κοιμάω, I sleep, I am calm

κοιλία, stomach, womb

κοινός, common, common assembly

κοινόω, I defile

κοινωνία, (related to κοινός, common) fellowship, participation

κοινωνός, partaker, partner, companion

κοπιάω, I toil, grow tired, am weary

κόπος, (from κόπτω, I labor) labor, trouble

κόσμος, world, mankind, earth, decoration, adornment (cosmos)

κράβαττος, bed, cot

κράζω, I call out

κρατέω, I grasp, am strong, take possession

κράτος, power, might

κρείττων, better

κρίμα, (from κρίνω [I judge]) judgment, decree, decision (critic)

κρίνω, I judge

κρίσις, (from κρίνω, I judge) judgment, decision (critic)

κριτής, (from κρίνω, I judge) judge (critic)

κρυπτός, hidden

κρύπτω, I hide

κτίζω, I create, build, found

κτίσις (from κτίζω, I create) creation, creature

κύριος, lord, master, the Lord

κώμη, village (commune)

κωλύω, I hinder

κωφός, deaf, unable to speak

λ

λαμβάνω, I take, receive, choose

λαλέω, I speak

λαός, people

λατρεύω, I serve, worship

λέγω, I say, speak, tell

λευκός, white

ληστής, thief
λίμνη, lake
λιμός, famine
λίαν, exceedingly
λίθος, stone (lithosphere)
λογίζομαι, I count, think, calculate
λόγος, (from λέγω, I say) word, speech, message, book, volume (logic)
λοιπός, rest, remaining, other
λοιπός, rest, remaining, other
λυπέω, I grieve, pain
λύπη, grief, pain
λυχνία, lampstand, candlestick
λύχνος, lamp
λύω, I loosen, release

μ

μαθητής, disciple, student. Literally = learner
μακάριος, blessed, happy, fortunate
μακράν, far, far off, far away
μακρόθεν, far off, from afar
μακροθυμία, patience, endurance
μάλιστα, especially, exceedingly
μανθάνω, I learn
μαρτυρέω, I bear witness, suffer martyrdom
μαρτυρία, (μάρτυς) testimony, martyrdom
μαρτύριον, testimony, martyrdom
μάρτυς, witness, martyr
μάχαιρα, dagger, short-sword
μέλος, body part, musical part, melody
μένω, I remain, abide, stay
μεριμνάω, I am anxious for
μερίζω, I divide, distribute

μέρος, part
μέσος, middle, (prep +gen) in the middle, (adverb) among
μετανοέω, I repent
μετάνοια, repentance
μέτρον, measurement, dimension, poetic meter
μή, not, lest
μηδέ, nor, and not
μηκέτι, no longer, no more
μήν, month
μήποτε, lest, perhaps
μήτε, and not, neither . . . nor
μήτηρ, mother (maternal)
μιμνῄσκομαι, I remember, remind
μικρός, small, little
μισέω, I hate
μισθός, wages, reward
μνημεῖον, tomb, monument
μνημονεύω (μιμνῄσκομαι), I remember, mention
μοιχεύω, I commit adultery
μόνος, only, alone
μόνος, only, alone
μύρον, ointment (myrrh)
μυστήριον, mystery, secret
μωρός, foolish (moron)

ν

ναός, temple, shrine, palace
νεανίσκος, (related to νέος, young) young man
νέος, new, young (a neologism is a newly coined word)
νεφέλη, cloud
νήπιος, child (nipple)
νηστεύω, I fast
νικάω, I conquer (Nike was the Greek goddess of victory)

νίπτω, I wash

νομίζω, I think, suppose

νόμος, law, principle (nomology is the science or study of laws)

νοέω, I understand

νόσος, disease

νοῦς, mind, thought

νυμφίος, bridegroom

νῦν, now, present

νυνί, (synonym of νῦν) now

νύξ, night

ξ

ξένος, stranger, strange (xenophobia)

ξηραίνω, I dry up

ξύλον, wood, tree

ο

ὀμνύω, I swear

ὁμοθυμαδόν, of one accord

ὅμοιος, like, similar

ὁμοιόω, I make like

ὁμοίως, likewise

ὁμολογέω, I confess, praise

ὁδός, road, way (odometer)

ὀδούς, tooth

οἶδα, I know

οἰκία, (synonym of οἶκος) house

οἰκοδεσπότης, householder

οἰκοδομέω, I build

οἰκοδομή, building

οἰκονόμος, manager, steward

οἶκος, house (economics)

οἰκουμένη, world
οἶνος, wine
ὀλίγος, little, few
ὄνομα, name
ὄντως, really, indeed
ὅπου, where, whereas
ὅπως, so that, in order that, how that
ὅραμα, (from ὁράω, I see) vision, sight
ὁράω, I see, view, perceive (oracle)
ὀργή, wrath, anger
ὅριον, border, coast, domain, limit
ὅρκος, oath
ὄρος, mountain (orology is the study of mountains)
ὅταν, when, whenever
ὅτε, when, while
ὅτι, that, because, since
οὗ, where, to where
οὐδέ, and not, neither, nor, not even
οὐδέ, and not, neither, nor, not even
οὐδέποτε, never
οὐκέτι, no longer
οὖν, therefore, thus
οὔπω, not yet
οὐρανός, heaven (Uranus)
οὖς, ear
οὔτε, neither
οὕτως, thus, so, in this way
ὀφείλω, I am obligated
ὀφθαλμός, (from ὁράω, I see) eye (ophthalmic)
ὄφις, snake
ὄχλος, crowd, multitude
ὀψία, evening

π

πάθημα, (from πάσχω, I suffer) suffering

παιδίον, child (pediatric)

παιδίσκη, young woman, servant (female)

παῖς, child, slave (pedagogy)

παλαιός, old (paleontology)

πάλιν, again, in so far as

παντοκράτωρ, all mighty, almighty

πάντοτε, always

παραβολή, proverb, poem, parable

παραγίνομαι, I come, appear

παραδίδωμι, I deliver, deliver over

παράδοσι, (from παρά [beside] + δίδωμι, [I give]) deliverance, tradition

παρακαλέω, (παρά [alongside] + καλέω [I call]) I urge, exhort, comfort

παράκλησις, (from παρά [from alongside] + καλέω [I call]) encouragement

παραλαμβάνω, I take

παραλυτικός, paralytic

παράπτωμα, (from παρά [beside], πίπτω [I fall]) trespass, sin

παρατίθημι, I set before, commend

παραχρῆμα, immediately

παρεμβολή, barracks, encampment

πάρειμι, I am present

παρέρχομαι, I pass by, pass away

παρέχω, I provide, cause, grant

παρθένος, virgin

παρίστημι, I present, stand by

παρουσία, coming, presence

παρρησία, boldness

πάσχα, (πάσχω) passover, passover lamb

πάσχω, I suffer

πατήρ, father, Father (patriarch)

παύω, I cease, stop

πέμπω, I send

πείθω, I persuade, believe, trust

πεινάω, I am hungry

πειράζω, I tempt, test, try

πειρασμός, temptation, test (pirate)

πέντε, five

περιπατέω, I walk, walk around (peripatetic = traveling from place to place)

περισσεύω, I abound

περισσός, more, remaining, excessive, to the full

περισσοτέρως, even more so

περιστερά, dove, pigeon

περιτέμνω, I circumcise

περιτομή, circumcision

πετεινόν, bird

πέτρα, rock (petrified)

πηγή, water spring

πίμπλημι, I fill, fulfill

πίνω, I drink

πίπτω, I fall

πιστεύω, I believe, trust

πίστις, (from πείθω, I believe) faith, belief, trust

πιστός, faithful, reliable, trustworthy

πλανάω (πλάνη), I deceive

πλάνη, error

πλεονεξία, greediness, lust

πληγή, plague, blow, wound

πλῆθος, multitude (plural)

πλήν, (+gen) only, except (prep.), but, nevertheless (conj.)

πλήρης, full

πληρόω, I fill, fulfill

πλήρωμα, fullness

πλησίον, (+gen) near, neighbor

πλοῖον, boat, ship

πλούσιος, rich

πλοῦτος, riches (plutarchy)

πνεῦμα, wind, breath, spirit, Spirit (pneumatic)

πνευματικός, spiritual

πόθεν, from where?

ποιμήν, shepherd

ποιέω, I do, make, work (poet)

ποικίλος, many colored, various kinds

πόλεμος, war (polemic)

πόλις, city, town (police)

πολύς, much, many

πονηρός, wicked, evil

πορεύομαι, I go

πορνεία, sexual immorality (pornography)

πόρνη, prostitute (pornography)

πόρνος, sexually immoral man, prostitute chaser (pornography)

ποταμός, river

ποτήριον, (from πίνω, I drink) cup

ποτίζω, I give a drink

πούς, foot

πρᾶγμα, (from πράσσω, I do) thing, deed

πράσσω, I do, accomplish

πραΰτης, gentleness

πρεσβύτερος, older, elder

πρίν, before, formerly

προάγω, I go before, elevate

πρόβατον, (from πρό [ahead] + βαίνω, [I go]) sheep

πρόθεσις,(πρό, τίθημι) setting forth, offering, purpose

προλέγω, I say ahead of time, foretell
προσδέχομαι, I receive, wait for
προσδοκάω, I wait for, expect
προσέρχομαι, I come, approach
προσευχή, prayer
προσεύχομαι, I pray
προσέχω, I pay attention to
προσκαλέω, I call, summon
προσκυνέω, I fall down, worship, do obeisance
προστίθημι, I put, add
προσφέρω, I bring, offer
πρόσωπον, (from πρός [to] + ὁράω [I see]) face
πρότερος, former, formerly
προφητεία, (from πρό, [before] + φημί [I say]) prophecy
προφητεύω, I prophesy
προφήτης, prophet
πρωΐ́, in the morning
πρῶτος, first, prominent, former
πτωχός, poor
πύλη, gate
πυλών, gate, porch
πῦρ, fire (pyre)
πωλέω, I sell
πῶλος, colt

ρ

ῥάβδος, stick, staff, scepter
ῥῆμα, word, thing
ῥίζα, root, origin
ῥύομαι, I deliver
Ῥωμαῖος, Roman

σ

σάββατον, sabbath

σαλεύω, I shake

σάλπιγξ, (from σαλπίζω, I sound the trumpet) trumpet

σάρξ, flesh, meat, body, sinful nature (a sarcophagus is a limestone coffin. This Greek word means "flesh-eating.")

σεισμός, (from σείω, I shake) earthquake, shakedown (extortion)

σημεῖον, sign, miracle

σήμερον, today

σῖτος, grain

σκανδαλίζω, I cause somebody to sin

σκάνδαλον, offense, cause of sin

σκεῦος, object, vessel

σκηνή, tent, Tabernacle

σκοτία, darkness

σκότος, darkness

σός, your (singular)

σοφία, skill, wisdom, philosophy

σοφός, wise, skilled

σπείρω, I sow

σπέρμα, (from σπείρω, I sow) seed (sperm)

σπλάγχνον, entrails, heart

σπουδή, (from σπεύδω, I hasten) haste, speed, zeal, pursuit, exertion

σταυρός, (from σταυρόω, I crucify) cross

σταυρόω, I crucify

στέφανος, crown

στόμα, mouth (stomach)

στρατηγός, captain, commander, chief magistrate

στρατιώτης, soldier

στρέφω, I turn

συμφέρω, I am better, bring together

συκῆ, fig tree

συνάγω, I gather, bring together

συναγωγή, (from σύν [with] + ἄγω [I gather]) collection, gathering, synagogue

σύνδουλος, (from σύν [with] + δοῦλος [slave]) fellow slave

συνείδησις, conscience

συνεργός, fellow worker

συνέρχομαι, I come together, go together

συνίστημι, I associate with, recommend, unite, collect

συνλαμβάνω, I seize

σφόδρα, very much

σφραγίζω, I seal

σφραγίς, seal

σῶμα, body (psychosomatic)

σῴζω, I save

σωτήρ, (from σῴζω, I save) savior, Savior

σωτηρία, (σῴζω) salvation

τ

τάλαντον, talent

ταράσσω, I trouble

ταχέως, quickly, soon

ταχύς, swift, quickly, soon

τέ, and, so, both/and

τέκνον, (from τίκτω, I give birth) (technology)

τέλειος, complete, perfect

τελειόω, I finish, make perfect

τελέω, I finish

τέλος, end, goal, tribute

τελώνης, tax collector

τέρας, wonder, marvel, omen

τέσσαρες, four

τεσσεράκοντα, forty

τηρέω, I keep, I guard
τιμάω, I honor
τιμή, honor
τίμιος, precious, rare
τίθημι, I put, make, appoint
τίκτω, I bear children
τολμάω, I dare, act rashly
τόπος, place, position, opportunity, passage (topography)
τότε, then
τράπεζα, table
τρεῖς, three
τρέχω, I run
τριάκοντα, thirty
τρίς, three times
τρόπος, way, manner, customs, kind of life, deportment
τροφή, (from τρέφω, I eat) food, provisions
τύπος, image, pattern, example, sign
τυφλός, blind

υ

ὑμέτερος, your (plural)
ὑγιής, whole, healthy
ὕδωρ, water (hydroelectric)
υἱός, son
ὑπάγω, I go away
ὑπακοή, (from ὑπό [under] + ἀκούω [I hear]) obedience
ὑπακούω, I obey, hear, listen
ὑπάρχω, I am, exist, possess, have advantage
ὑπηρέτης, assistant
ὑπομένω, I endure, remain, wait upon
ὑπομονή, endurance, staying
ὑπόδημα, sandal, shoe
ὑποκριτής, hypocrite

ὑποστρέφω, I return
ὑποτάσσω, I subject, submit
ὑστερέω, I lack, am late, postpone
ὕστερος, afterwards, last, later
ὑψηλός, high, proud
ὕψιστος, highest, most high, Most High
ὑψόω, I lift up, I exalt

Φ

φαίνω, I appear, shine, cause to shine
φανερός, known, plain
φανερόω, (synonym of φαίνω), I make known, reveal
φέρω, I bring, carry, (passive, I am extant)
φεύγω, I flee
φημί, I say
φιάλη, bowl
φιλέω, I love, have affection for, kiss
φίλος, friend, beloved, friendly, dear, pleasant, welcome
φοβέω, (related to φόβος [fear]), I fear (phobia)
φόβος, fear, terror, reverence (phobia)
φρονέω, I am wise, think
φρόνιμος, wise
φυλακή, (from φυλάσσω, I guard) guard, watch, prison (prophylactic)
φυλάσσω, I guard, keep
φυλή, tribe
φύσις, nature
φωνέω, (related to φωνή, [voice]), I call, shout
φωνή, voice, sound, speech (phone)
φῶς,(φαίνω, I shine) light (photograph)

Χ

χαίρω, I rejoice, hail

χαρά, joy
χαρίζομαι, I give, favor, forgive
χάρις, grace, favor, gratitude, gift
χάρισμα, gift (charismatic)
χείρ, hand (chiropractor)
χείρων, worse
χήρα, widow
χιλίαρχος, commander of a thousand, tribune
χιλιάς, group of a thousand
χίλιοι, thousand
χιτών, tunic, shirt
χοῖρος, pig
χορτάζω, I feed, fill
χόρτος, grass (chloroform)
χρεία, (from χράομαι, I need)) need, use, duty
χρηστότης, kindness
Χριστός, Christ
χρόνος, time (chronology)
χρυσίον, (synonym of χρυσός) gold, golden
χρυσός, gold
χρυσοῦς, golden
χωλός, crippled
χώρα, place, land, country
χωρίον, place, field

ψ

ψευδοπροφήτης, false prophet
ψεῦδος, (from ψεύδομαι, I lie) lie, falsehood
ψεύστης, (from ψεύδομαι, I lie) liar
ψυχή, soul (psyche)

ω

ὧδε, here, in this way, so, thus

ὥρα, hour
ὡς, as, like, while, that
ὡσαύτως, likewise
ὥσπερ, as, just as
ὥστε, so that, in order that, thus
ὠφελέω, I gain, benefit

BIBLIOGRAPHY

Archibald, Zofia. *Ancient Greeks.* 10th ed. New York, NY. Quarto Publishers, plc, 1991.

Arnold, Clinton E. *Exegetical Commentary on the New Testament.* 1st ePub ed. Grand Rapids, MI. Zondervan Publishing, 2016.

Atland. Kurt. *The Greek New Testament.* 3rd ed. Stuttgart, Germany. Biblia-Druck GmbH, 1983.

Barclay, William, *The Gospel of Mark.* 3rd ed. Philadelphia, PA. 1975.

Barker, Kenneth. *The NIV Study Bible.* 12th ed. Grand Rapids, MI. Zondervan Bible Publishers, 1985.

Baugh, S.M. *A First John Reader.* 1st ed. Phillipsburg, NJ. P & R Publishing, 1999.

Baugh, S.M. *A New Testament Greek Primer.* 1st ed. Phillipsburg, NJ. P & R Publishing, 1995.

Beale, G.K. *An Interpretive Lexicon of New Testament Greek.* 1st Kindle ed. Grand Rapids MI, Zondervan Publishing House, 2014.

Berry, George Ricker. *The Classic Greek Dictionary.* 7th ed. Chicago, IL. Follett Publishing Company, 1962.

Blass, F. *A Greek Grammar of the New Testament.* 10th ed. Chicago. The University of Chicago Press, 1973.

Bowker, John. *The Complete Bible Handbook.* 1st ed. New York, NY. DK Publishing, 1998.

Brenton, Lancelot. *The Septuagint with Apocrypha: Greek and English.* 11th ed. Peabody, MA. Hendrickson Publishers, 2005.

Brunschwig, Jacques. *Greek Thought.* 1st ed. Cambridge, MA. The Belknap Press of Harvard University Press, 2000.

Carr, A. *St. Matthew.* 12th ed. Cambridge, MA. Cambridge University Press, 1913.

Cartledge, Paul. *Ancient Greece.* 1st ed. Cambridge, MA. Cambridge University Press, 1998.

Comrie, Bernard. *The World's Major Languages.* 3rd ed. New York. Routledge, 2018.

Cruden, Alexander. *Cruden's Complete Concordance.* 8th ed. Grand Rapids, MI. Zondervan Publishing House, 1968.

Dana, Julius R. *A Manual Grammar of the Greek New Testament.* 2nd ed. New Yor. The Macmillan Company, 1962.

Decker, Rodney J. *Reading Koine Greek.* 1st ed. Grand Rapids. MI. Baker Academic, 2014.

Dover, Kenneth. *The Greeks.* 2nd ed. Austin, TX. University of Texas Press, 1981.

Drews, Robert. *The Coming of the Greeks.* 2nd ed. Princeton, New Jersey. Princeton University Press, 1988.

Douglas, J.D. *New Bible Dictionary.* 2nd ed. Wheaton, IL: Tyndale House Publishers, Inc, 1984.

Finley, M.I. *The Ancient Greeks.* 2nd ed. New York. Penguin Books, 1964.

Found, Jim. Basic Greek in 30 Minutes a Day. 1st ed. Minneapolis, MN. Bethany House Publishers, 1983.

Friederich, Werner p. *An Outline History of German Literature.* 3rd ed. New York, NY. Barnes and Noble, Inc. 1966.

Gleaves, Scott G. *Did Jesus Speak Greek?* 1st eBook ed. Eugene, OR. Pickwick Publications, 2015.

Gower, Ralph. *The New Manners and Customs of Bible Times.* 3rd ed. Oxford, England. Moody Publishers, 2005.

Green, Peter. *Alexander to Actium.* 2nd ed. Berkeley, CA. University of California Press, 1993.

Hammond, N.G.L. *A History of Greece.* 2nd ed. Oxford, England. Oxford University Press, 1963.

Hewett, James Allen. *New Testament Greek.* 4th ed. Peabody, MA. Hendrickson Publishers, Inc., 1986.

Köstenberger, Andreas J. *Going Deeper Greek.* 2nd ed. Nashville, TN. B & H Academics, 2020.

LaSor, William Sanford. *Handbook of New Testament Greek, Volume One.* 2nd ed. Grand Rapids, MI: Williams B. Eerdmans Publishing Company, 1974.

LaSor, William Sanford. *Handbook of New Testament Greek. Volume Two.* 2nd ed. Grand Rapids, MI: Williams B. Eerdmans Publishing Company, 1974.

Lightfoot, Neil R. *How We Got The Bible.* 3rd ed. Kindle Version. Grand Rapids, MI. Baker Books, 2015.

Machen, J. Gresham. *New Testament Greek for Beginners.* 1st ed. Toronto, Canada. The Macmillan Company, 1923.

Macy, John. *The Story of the World's Literature.* 2nd ed. Liveright, New York. Liveright Publishing Corp., 1936.

Manchester, William. *A World Only Lit by Fire.* 1st ed. Boston, MA. Little, Brown and Company, 1993.

Marshall, Alfred. *The Interlinear Greek-English New Testament.* 1st ed. Grand Rapid. MI. Zondervan Publishing House, 1975.

Martin, Thomas R. *Ancient Greece.* 1st ed. Ann Arbor, MI. Edwards Brothers, Inc., 1996.

Matyszak, Philip. *The Greeks.* 1st ed. London. Reaktion Books, 2018.

Merkle, Benjamin L. *1 John.* 1st ed. Brentwood, TN. B & H Academic, 2024.

Miller, Stephen M. *The Bible: A History.* 1st ed. Intercourse, PA. Good Books, 2003.

Mounce, William D. *Basics of Biblical Greek.* 2nd ed. Grand Rapids, MI. Zondervan Publishers, 2003.

Murry, Oswyn. *Early Greece.* 2nd ed. Cambridge, MA. Harvard University Press, 1993.

Ober, Josiah. *The Rise and Fall of Classical Greece.* 1st ed. Princeton, New Jersey. Princeton University Press, 2015.

O,Neill, Amanda. *Biblical Times.* 2nd ed. Baltimore, MD. Ottenheimer Publishers, Inc, 1995.

Pennington, Jonathan T. *The Sermon on the Mount.* 1st eBook ed. Grand Rapids, MI. Baker Publishing Group, 2017.

Peters, F.E. *The Harvest of Hellenism.* 1st ed. New York. Barnes and Noble, 1996.

Phillips, John. *Exploring Hebrews.* 3rd ed. Grand Rapids, MI. Kregel Publications, 2002.

Ramey, William. "NT Greek in Session." Why Study Greek?, accessed January 5, 2026, https://www.inthebeginning.org/ntgreek/introduction.html.

Robertson, A.T. *A Grammar of the Greek New Testament in the Light of Historical Research.* 4th ed. Nashville, TN. Broadman Press, 1934.

Robinson, Maurice A. *The New Testament in the Original Greek.* 1st ePub ed. Cambridge, MA. Anchor-Cross Publishing, 2005.

Robinson, Thomas A. *Mastering Greek Vocabulary.* 2nd ed. Peabody, MA. Hendrickson Publishers, Inc., 1991.

Rösch, Konstantin. *Das Neue Testament.* 1st ed., Munich. Verlag Ferdinand Schöningh, 1967.

Ryrie, Charles C. *Biblical Theology of the New Testament.* 4th ePub ed. Dubuque, Ia. ESC Ministries, 2005.

Schoder, Raymond V. *A Reading Course in Homeric Greek.* 2nd ed. Chicago, IL. Loyola University Press, 1985.

Scofield, C.I. *Die Heilige Schrift.* 3rd ed. Westchester, IL. Good News Publishers, 1972.

Tenney, Merrill C. *A Parsing Guide to the Greek New Testament.* 1st ed. Scottsdale, PA: Herald Press, 1971.

Thompson, Frank Charles. *The NKJV Thompson Chain Bible.* 2nd eBook ed. Grand Rapids, MI, Zondervan Publishing, 2023.

Turner, Nigel. *Christian Words.* 1st ed. Nashville, Thomas Nelson Publishers, 1982.

Van Doren, Charles. *A History of Knowledge.* 1st ed. New York, Ballantine Books, 1991.

Wallace, Daniel B. *The Basics of New Testament Greek Syntax.* 1st ed. Kindle Edition. Grand Rapids, MI. Zondervan Publishers, 2000.

Weatherly, Edward H. *The Heritage of European Literature, Volume One.* 1st ed. Boston, Ginn and Company, 1948.

Wedeck, Harry E. *Classics of Greek Literature.* 1st ed. New York. Philosophical Library, Inc., 1963.

Whitacre, Rodney A. *A Grammar of the Greek New Testament.* 1st ed. Grand Rapids, MI. Eerdmans Publishing Company, 2021.

Zodhiates, Spiros. *Learn or Review New Testament Greek.* 1st ed. Chattanooga, TN. AMG Publishers, 1976.

Made in United States
Troutdale, OR
11/03/2026

56847396R00156